AF356132

Practical Applications of NMR to Solve Real-World Problems

Practical Applications of NMR to Solve Real-World Problems

Editor

Robert Brinson

MDPI • Basel • Beijing • Wuhan • Barcelona • Belgrade • Manchester • Tokyo • Cluj • Tianjin

Editor
Robert Brinson
National Institute of Standards and Technology and the University of Maryland
USA

Editorial Office
MDPI
St. Alban-Anlage 66
4052 Basel, Switzerland

This is a reprint of articles from the Special Issue published online in the open access journal *Molecules* (ISSN 1420-3049) (available at: https://www.mdpi.com/journal/molecules/special_issues/Practical_NMR).

For citation purposes, cite each article independently as indicated on the article page online and as indicated below:

LastName, A.A.; LastName, B.B.; LastName, C.C. Article Title. *Journal Name* **Year**, *Volume Number*, Page Range.

ISBN 978-3-0365-2580-8 (Hbk)
ISBN 978-3-0365-2581-5 (PDF)

Contents

molecules

MDPI

Editorial

Practical Applications of NMR to Solve Real-World Problems

Robert G. Brinson

Institute for Bioscience and Biotechnology Research, National Institute of Standards and Technology and The University of Maryland, 9600 Gudelsky Drive, Rockville, MD 20850, USA; robert.brinson@nist.gov

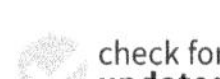

check for **updates**

Citation: Brinson, R.G. Practical Applications of NMR to Solve Real-World Problems. *Molecules* **2021**, *26*, 7091. https://doi.org/10.3390/molecules26237091

Received: 10 November 2021
Accepted: 22 November 2021
Published: 24 November 2021

Publisher's Note: MDPI stays neutral with regard to jurisdictional claims in published maps and institutional affiliations.

Nuclear magnetic resonance spectroscopy (NMR) is known to be a powerful technique for the characterization of small molecules and structural and dynamics studies of biomolecules. While it was once primarily used only in academia, this technique is now routinely utilized in many industry sectors, government agencies, and other applied research activities with great benefits. Such areas include agriculture, metabolomics and complex mixtures, manufacturing and process monitoring, pharmaceuticals (biologics and small molecules), national security, forensics, energy, and renewables. Many similar technical challenges, and often, regulatory burdens, are shared across these fields; yet, there is often little cross-talk between the various NMR practitioners in each of these focused areas. Greater sharing of experience and expertise amongst practitioners of applied NMR would be of benefit to all.

With this in mind, this Special Issue collected 10 original papers covering four sectors, including the Oil Industry, Nanostructured Systems and Materials, Metabolomics, and Biologics. These contributions show how this technology is now used in very disparate areas. Further, the NMR systems discussed range from low-field magnetic resonance imaging (MRI), 18.2 MHz relaxometry, and high fields up to 700 MHz. These articles show that NMR technology can be tailored for the specific needs of an industry.

Oil Industry: In their work entitled *"Study on Nuclear Magnetic Resonance Logging T_2 Spectrum Shape Correction of Sandstone Reservoirs in Oil-Based Mud Wells"* Sun et al. presented a need to improve the NMR logging data in order to accurately determine the physical parameters of the surrounding sandstone reservoirs to improve drilling operations [1]. During drilling operations, overbalanced pressure causes the infiltration of oil-based drilling lubricant into the surrounding sediment, resulting in a shorter T_2 relaxation. The authors determined a T_2 correction factor so that accurate NMR logging data could be determined for the morphology of the sandstone reservoirs.

In their article entitled *"Multiphase Flow Regime Characterization and Liquid Flow Measurement Using Low-Field Magnetic Resonance Imaging"*, Tromp and Cerioni simulated a multiphase flow in a pipeline [2]. They used low-field magnetic resonance imaging (MRI) to measure various types of flow and found the MRI measurement offered an accurate flow determination that was not dependent upon the multiphasic character of the flow. They suggest that the MRI measurement technology is robust enough for routine implementation in the petroleum industry.

Nanostructured Systems and Materials: In an article by Shelyapina et al., entitled *"^1H NMR Study of the $HCa_2Nb_3O_{10}$ Photocatalyst with Different Hydration Levels"*, ^{1}H magic angle spinning (MAS) NMR and ^{1}H spin-lattice relaxation time in the rotating frame $(T_{1\rho})$ were used to investigate the hydration properties of a layered perovskite-like oxide, $HCa_2Nb_3O_{10}$ [3]. This specific perovskite-like oxide is known to have photocatalytic properties, and the state of the water molecules influences these properties. From their ^{1}H NMR studies, the authors determined that $HCa_2Nb_3O_{10}$ has three different hydration forms: α-form $HCa_2Nb_3O_{10}\cdot1.6H_2O$, β-form $HCa_2Nb_3O_{10}\cdot0.8H_2O$, and γ-form $HCa_2Nb_3O_{10}\cdot0.1H_2O$.

In an effort to improve separation and purification protocols, a new method was introduced to study the solvent separation efficiency of metal–organic frameworks (MOFs) [4].

As described by Wagemann et al. in *"Screening Metal–Organic Frameworks for Separation of Binary Solvent Mixtures by Compact NMR Relaxometry"*, the measurement of effective transverse relaxation by low-field ^{1}H NMR relaxometry of a two-solvent mixture was used to establish correlation curves in relation to their mass proportion. Using the model MOF, powdered UiO-66 (Zr), the authors demonstrated great accuracy with their method and agreement with solution-state ^{1}H NMR measurements. They suggest their method could be integrated into the workflow of a synthetic laboratory due to their method's accuracy, ability for automation, and time savings.

Metabolomics: Honrao and coworkers presented paramagnetic relaxation enhancement (PRE) to increase sensitivity for metabolomics measurements in *"Gadolinium-Based Paramagnetic Relaxation Enhancement Agent Enhances Sensitivity for NUS Multidimensional NMR-Based Metabolomics"* [5]. By doping reference and test metabolite samples with gadolinium, the authors demonstrated a sensitivity enhancement, especially for the weakest signals, resulting in a lower limit of detection (LOD) and limit of quantification (LOQ). Further, their method maintains the linearity of intensity needed for the concentration range of metabolites tested.

The article *"Establishing a Metabolite Extraction Method to Study the Metabolome of Blastocystis Using NMR"* by Newton et al. describes a method of extracting metabolites from *Blastocytis* [6]. More in-depth analysis of the role of metabolism is needed to understand this pathogenic human parasite. The researchers tested a variety of extraction conditions, including solvent, lysis technique, and temperature. They optimized the first published extraction protocol for NMR analysis of the *Blastocystis* metabolome.

A method to combine both NMR and high-resolution mass spectrometry (HRMS) was proposed by Petrella et al. in *"Personalized Metabolic Profile by Synergic Use of NMR and HRMS"* [7]. Their SYNHMET (SYnergic use of NMR and HRMS for METabolomics) protocol afforded a synergistic characterization of urine metabolites, leading to higher accuracy in identification and concentration determination. In all, this new method detected 165 metabolites from patients with varying health profiles.

Biologics: In an article entitled *"Glycosylation States on Intact Proteins Determined by NMR Spectroscopy"*, Hargett et al. proposed an NMR method to selectively analyze the glycans from intact glycoproteins using RNase A and RNase B as model proteins [8]. Specifically, they implemented a ^{1}H,^{13}C HSQC-TOCSY at natural isotopic abundance. By optimizing the TOCSY mixing period, they could suppress the protein NMR signals, allowing for the analysis of the glycans independent of the protein. They plan to apply their work to polysaccharide conjugate vaccines.

The Wikström laboratory performed a comparative analysis of methods to evaluate higher-order structures (HOS) of monoclonal antibody-based therapeutics in *"Use of the 2D ^{1}H-^{13}C HSQC NMR Methyl Region to Evaluate the Higher Order Structural Integrity of Biopharmaceuticals"* [9]. Comparing near-ultraviolet circular dichroism (NUV-CD), intrinsic fluorescence spectroscopy (FLD), and natural abundance ^{1}H,^{13}C HSQC experiments, the low-resolution methods only provided limited discrimination between folded and unfolded samples of IgG1 and IgG2 subtypes. However, the NMR HOS method demonstrated high sensitivity to even subtle changes in higher-order structures (HOS) and may serve as a replacement method for low-resolution HOS assessments.

Another study, entitled *"NMR Spectroscopy for Protein Higher Order Structure Similarity Assessment in Formulated Drug Products"* by Wang et al., looked at defining similarity metrics for protein-based drug products (DPs) [10]. Using several different peptide and proteins, they presented a variety of metrics for establishing the similarity between a biosimilar DPs. Their overall goal was to establish a fit-for-purpose quantitative metric to establish similarity for a biosimilar.

As a guest editor of this Special Issue, I would like to extend my appreciation to all the authors for their excellent contributions and for their valuable support. I additionally thank the reviewers for their time and feedback.

Funding: This research received no external funding.

Conflicts of Interest: The author declares no conflict of interest.

Disclaimer: Certain commercial equipment, instruments, and materials are identified in this paper in order to specify the experimental procedure. Such identification does not imply recommendation or endorsement by the National Institute of Standards and Technology, nor does it imply that the material or equipment identified is necessarily the best available for the purpose.

References

1. Sun, J.; Cai, J.; Feng, P.; Sun, F.; Li, J.; Lu, J.; Yan, W. Study on Nuclear Magnetic Resonance Logging T2 Spectrum Shape Correction of Sandstone Reservoirs in Oil-Based Mud Wells. *Molecules* **2021**, *26*, 6082. [CrossRef] [PubMed]
2. Tromp, R.R.; Cerioni, L.M.C. Multiphase Flow Regime Characterization and Liquid Flow Measurement Using Low-Field Magnetic Resonance Imaging. *Molecules* **2021**, *26*, 3349. [CrossRef] [PubMed]
3. Shelyapina, M.G.; Silyukov, O.I.; Andronova, E.A.; Nefedov, D.Y.; Antonenko, A.O.; Missyul, A.; Kurnosenko, S.A.; Zvereva, I.A. ^{1}H NMR Study of the $HCa_2Nb_3O_{10}$ Photocatalyst with Different Hydration Levels. *Molecules* **2021**, *26*, 5943. [CrossRef] [PubMed]
4. Wagemann, M.; Radzik, N.; Krzyzak, A.; Adams, A. Screening Metal-Organic Frameworks for Separation of Binary Solvent Mixtures by Compact NMR Relaxometry. *Molecules* **2021**, *26*, 3481. [CrossRef] [PubMed]
5. Honrao, C.; Teissier, N.; Zhang, B.; Powers, R.; O'Day, E.M. Gadolinium-Based Paramagnetic Relaxation Enhancement Agent Enhances Sensitivity for NUS Multidimensional NMR-Based Metabolomics. *Molecules* **2021**, *26*, 5115. [CrossRef] [PubMed]
6. Newton, J.M.; Betts, E.L.; Yiangou, L.; Ortega Roldan, J.; Tsaousis, A.D.; Thompson, G.S. Establishing a Metabolite Extraction Method to Study the Metabolome of Blastocystis Using NMR. *Molecules* **2021**, *26*, 3285. [CrossRef] [PubMed]
7. Petrella, G.; Montesano, C.; Lentini, S.; Ciufolini, G.; Vanni, D.; Speziale, R.; Salonia, A.; Montorsi, F.; Summa, V.; Vago, R.; et al. Personalized Metabolic Profile by Synergic Use of NMR and HRMS. *Molecules* **2021**, *26*, 4167. [CrossRef] [PubMed]
8. Hargett, A.A.; Marcella, A.M.; Yu, H.; Li, C.; Orwenyo, J.; Battistel, M.D.; Wang, L.X.; Freedberg, D.I. Glycosylation States on Intact Proteins Determined by NMR Spectroscopy. *Molecules* **2021**, *26*, 4308. [CrossRef] [PubMed]
9. Hwang, T.L.; Batabyal, D.; Knutson, N.; Wikstrom, M. Use of the 2D ^{1}H-^{13}C HSQC NMR Methyl Region to Evaluate the Higher Order Structural Integrity of Biopharmaceuticals. *Molecules* **2021**, *26*, 2714. [CrossRef] [PubMed]
10. Wang, D.; Zhuo, Y.; Karfunkle, M.; Patil, S.M.; Smith, C.J.; Keire, D.A.; Chen, K. NMR Spectroscopy for Protein Higher Order Structure Similarity Assessment in Formulated Drug Products. *Molecules* **2021**, *26*, 4251. [CrossRef] [PubMed]

Article

Study on Nuclear Magnetic Resonance Logging T_2 Spectrum Shape Correction of Sandstone Reservoirs in Oil-Based Mud Wells

Jianmeng Sun [1,2], Jun Cai [3], Ping Feng [2,*], Fujing Sun [2], Jun Li [1], Jing Lu [1] and Weichao Yan [2]

[1] State Energy Center for Shale Oil Research and Development, Beijing 100083, China; sunjm@upc.edu.cn (J.S.); lijun67.syky@sinopec.com (J.L.); lujing.syky@sinopec.com (J.L.)
[2] School of Geosciences, China University of Petroleum (East China), Qingdao 266580, China; b20010072@upc.edu.cn (F.S.); yanweichaoqz@163.com (W.Y.)
[3] Shanghai Branch of CNOOC Ltd., Shanghai 200030, China; caijun@cnooc.com.cn
* Correspondence: upcfengping@163.com; Tel.: +86-17863960551

Abstract: The oil-based mud filtrate will invade the formation under the overbalanced pressure during drilling operations. As a result, alterations will occur to the nuclear magnetic resonance (NMR) response characteristics of the original formation, causing the relaxation time of the NMR T_2 spectrum of the free fluid part to move towards a slower relaxation time. Consequently, the subsequent interpretation and petrophysical evaluation will be heavily impacted. Therefore, the actual measured T_2 spectrum needs to be corrected for invasion. For this reason, considering the low-porosity and low-permeability of sandstone gas formations in the East China Sea as the research object, a new method to correct the incorrect shape of the NMR logging T_2 spectrum was proposed in three main steps. First, the differences in the morphology of the NMR logging T_2 spectrum between oil-based mud wells and water-based mud wells in adjacent wells were analyzed based on the NMR relaxation mechanism. Second, rocks were divided into four categories according to the pore structure, and the NMR logging T_2 spectrum was extracted using the multidimensional matrix method to establish the T_2 spectrum of water-based mud wells and oil-based mud wells. Finally, the correctness of the method was verified by two T_2 spectrum correction examples of oil-based mud wells in the study area. The results show that the corrected NMR T_2 spectrum eliminates the influence of oil-based mud filtrate and improves the accuracy of NMR logging for calculating permeability.

Keywords: NMR; oil-based mud; invasion correction; permeability

Citation: Sun, J.; Cai, J.; Feng, P.; Sun, F.; Li, J.; Lu, J.; Yan, W. Study on Nuclear Magnetic Resonance Logging T_2 Spectrum Shape Correction of Sandstone Reservoirs in Oil-Based Mud Wells. *Molecules* **2021**, *26*, 6082. https://doi.org/10.3390/molecules26196082

Academic Editor: Robert Brinson

Received: 31 August 2021
Accepted: 2 October 2021
Published: 8 October 2021

Publisher's Note: MDPI stays neutral with regard to jurisdictional claims in published maps and institutional affiliations.

1. Introduction

During drilling operations, drilling fluids are primarily used to ensure the smooth progress of drilling, and their importance is equivalent to the blood of the human body. The drilling fluid can carry out drilled cuttings, cool and lubricate the bits, balance and control the formation pressure, and maintain the stability of the wellbore [1]. To ensure the safety of drilling operation, the drilling fluid is usually circulated under a bottom hole pressure higher than the reservoir pressure, which is called overbalanced drilling [2]. Therefore, a fraction of the drilling fluid filtrate will penetrate the permeable formation under overbalanced pressure. Then, it will interact with the minerals and fluids in the formation rock, which will change the original petrophysical and geomechanical properties in the formation flushing zone [3]. The degree of invasion usually depends on two factors. The internal factors mainly include the quality control of reservoir formation, such as porosity, permeability, pore structure, original wettability of rock, and formation fluid properties. External factors are primarily controlled by drilling operation parameters, including drilling pressure, formation temperature, drilling time and drilling fluid properties [4]. On the walls of the wellbore, the solid particles in the mud are gradually deposited to form mud cakes. Because mud cakes are impermeable, when mud cakes are formed, they hinder

5

the further invasion of mud filtrates [5]. As shown in Figure 1, this process will make the reservoir near the wellbore form several obvious annular areas: mud cake, invasion zone, undisturbed zone and impermeable zone.

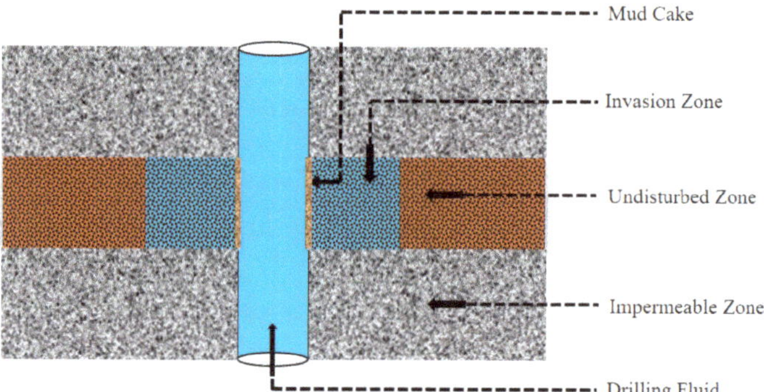

Figure 1. Schematic diagram of mud invasion process.

Water-based mud (WBM) and oil-based mud (OBM) are considered the two types of drilling fluids that are most commonly used in the drilling process. Because shale oil reservoirs are rich in clay minerals and rock stability is poor, drilling with water-based mud will lead to water absorption and swelling in the rock, hydration expansion, borehole collapse, leakage, and so on [6,7]. Oil-based mud can effectively solve these problems, such as hydration expansion and wellbore instability of shale formations. Therefore, oil-based mud is most commonly used in the drilling process of shale oil reservoirs [8].

Nuclear magnetic resonance logging can directly measure the relaxation information of reservoir pore fluids and plays an extremely important role in the evaluation of petro-physical parameters, such as porosity, permeability, and saturation of various types of oil and gas reservoirs [9,10]. Based on the theory of NMR logging, the NMR transverse relaxation time (T_2) is mainly affected by three relaxation mechanisms: bulk relaxation, surface relaxation, and diffusion relaxation, as shown in Equation (1):

$$\frac{1}{T_2} = \frac{1}{T_{2B}} + \rho_2 \frac{S}{V} + \frac{D(\gamma G T_E)^2}{12} \tag{1}$$

where T_2 is the transverse relaxation time, T_{2B} is the bulk relaxation, ρ_2 is the surface relaxivity, S/V is the surface-to-volume ratio of the pore, D is the diffusion coefficient of the fluid, γ is the nuclear gyromagnetic ratio, G is the magnetic field gradient, and T_E is the echo spacing of the measurement sequence.

However, due to the shallow detection depth of NMR logging, the T_2 spectrum is easily affected by mud invasion. Especially in the environment of oil-based drilling fluids in shale oil reservoirs, the displacement of fluid in the flushed zone by oil-based mud filtrate may seriously affect the overall shape of the T_2 spectrum and cannot accurately evaluate reservoir parameters [11].

The study area belongs to a typical low porosity and low permeability sandstone gas reservoir with water content (the original water saturation range is 23–85%) and strong hydrophilic characteristics [12]. In this context, the rock volume model of the gas reservoir and the corresponding ideal NMR T_2 spectrum were established, as shown in Figure 2. The model describes the ideal NMR T_2 spectrum of gas-bearing reservoirs in undisturbed formations of water-based mud wells and oil-based mud wells.

1. The undisturbed formation does not contain mud filtrate, and the natural gas is a fluid of nonwetting phase in the formation pores of the study area. Its NMR signal only includes diffusion relaxation and free relaxation, which is not affected

by surface relaxation. Due to the fast diffusion rate of natural gas, it has a short T_2 time [13]. Figure 2a shows the rock volume model and ideal NMR T_2 spectrum of the undisturbed formation, and the red part represents natural gas.

2. In the process of nuclear magnetic resonance logging, the response of the instrument will be affected by the mud filtrate in the invasion zone. Figure 2b shows the rock volume model and ideal nuclear magnetic resonance T_2 spectrum in the water-based mud well: some natural gas and movable water in the formation invasion zone in the water-based mud well are displaced by the water-based mud filtrate, and the properties of the fluid are changed. The free gas product in the intrusion zone is less, and the signal of gas in the corresponding NMR T_2 spectrum decreases. Because the wettability of the formation in the study area is strong water wet, the filtrate of water-based mud contacts the skeleton minerals on the pore surface, and its NMR response characteristics are affected by surface relaxation. Therefore, the T_2 signals of the water-based mud filtrate and movable water overlap. For sandstone gas reservoirs, researchers usually use the T_2 spectrum of nuclear magnetic resonance logging in water-based mud wells to calculate basic reservoir parameters.

3. Figure 2c shows the rock volume model and ideal nuclear magnetic resonance T_2 spectrum in the oil-based mud well: in the oil-based mud well, in addition to some natural gas and movable water in the formation invasion zone being displaced by the oil-based mud filtrate, the oil-based mud filtrate will also be miscible with the residual natural gas in the invasion zone, resulting in a significant reduction in the natural gas signal collected by nuclear magnetic resonance logging. Because the filtrate of oil-based mud is in a nonwetting phase, the flow channel in the formation rock is the central area of the pore and does not contact the pore surface. Its NMR response is mainly characterized by volume relaxation. The viscosity of oil-based mud in the study area is 18 mPa.s, which is a low viscosity drilling fluid. Its body relaxation time is long, which significantly changes the T_2 spectrum shape of the undisturbed formation. Therefore, if the existing reservoir parameter calculation methods are directly applied to the T_2 spectrum of NMR logging in oil-based mud wells, it may lead to large calculation errors.

Considering the character analysis of nuclear magnetic resonance logging data in oil-based mud wells, many researchers have conducted related research [14–17]. Chen [14,15] believed that surfactants in oil based drilling fluids would change the wettability of reservoir rocks, and thus using the default 33 ms as the $T_{2cut\text{-}off}$ value for interpretation is no longer applicable; Marschall and Coats [16] believe that the macropores reflected by the T_2 spectrum are greatly affected by the oil-based mud filtrate, while the small pores are not sensitive to the intrusion of the oil-based mud filtrate. However, the method of morphological correction of the T_2 spectrum of nuclear magnetic resonance logging in oil-based mud wells is not well studied. Ighodalo [17] proposed a fluid substitution method to correct the nuclear magnetic T_2 spectrum in oil-based mud wells, yet this method needs to accurately obtain the total water saturation in the invasion zone. Therefore, a method to quickly and accurately correct the shape of the NMR T_2 spectrum after the invasion of oil-based mud filtrate in shale oil reservoirs is needed. However, due to the complex pore structure of shale oil reservoirs and the complex response mechanism of NMR logging, correction work is difficult. In this regard, this study proposed a new approach to correct the unreal NMR logging responses due to oil-based mud invasion in gas sandstone formations. By establishing the multivariate linear function relationship between the nuclear magnetic T_2 spectrum of water-based mud wells and oil-based mud wells, the shape correction of the NMR logging T_2 spectrum after the invasion of oil-based mud filtrates are realized, which improved the accuracy of NMR logging evaluation of reservoir parameters, and laid a foundation for the shape correction of the nuclear magnetic resonance logging T_2 spectrum after the invasion of oil-based mud filtrates in shale oil reservoirs.

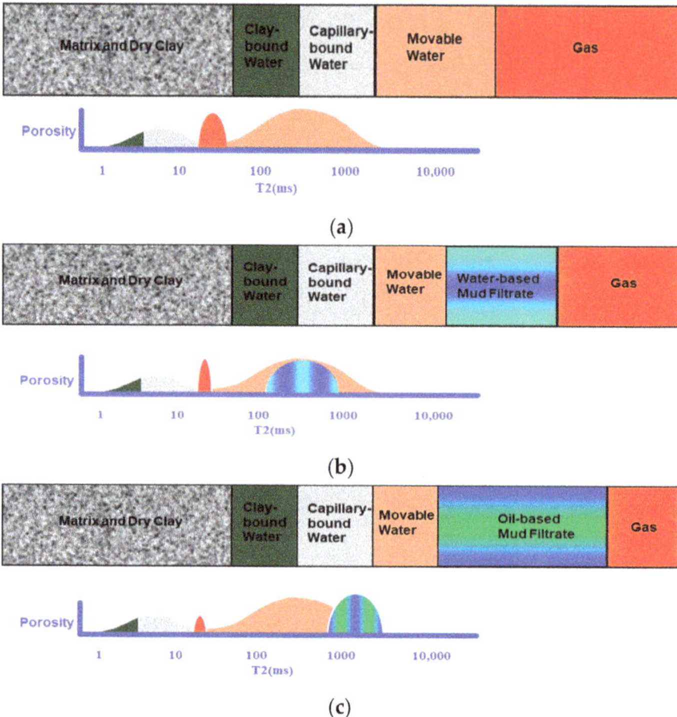

Figure 2. Gas reservoir volume model and theoretical NMR T_2 spectrum diagram. (**a**) Volume model of gas-bearing reservoirs in undisturbed formations and theoretical distribution of the nuclear magnetic T_2 spectrum. (**b**) Volume model of gas-bearing reservoirs in water-based mud wells and theoretical distribution of the nuclear magnetic T_2 spectrum. (**c**) Volume model of gas-bearing reservoirs in oil-based mud wells and theoretical distribution of the nuclear magnetic T_2 spectrum.

2. Results and Discussion

2.1. Comparison of NMR Logging Responses in Different Mud Environments

As shown in Figure 2, under the condition of drilling differential pressure, the original formation fluid was displaced by mud filtrate, and the properties of the formation fluid have changed. Moreover, the wettability of the formation determines the different distribution positions of oil-based mud filtrate and water-based mud filtrate in pores, resulting in the difference in the T_2 spectrum morphology of NMR logging under different drilling fluid environments.

In this study, the T_2 spectrum characteristics of water-based mud well A and adjacent oil-based mud well B were compared and analyzed. Figure 3a is a comprehensive logging diagram of well A measured in a water-based mud environment. The sandstone interval is a gas reservoir, and its NMR T_2 spectrum is distributed in a single peak. The T_2 value of the main peak is less than 300 ms, and the maximum T_2 value is less than 1000 ms. According to the volume model of gas-bearing reservoirs in water-based mud wells and the theoretical distribution of the nuclear magnetic T_2 spectrum in Figure 2b, the distribution form of the nuclear magnetic T_2 spectrum in water-based mud wells is basically not affected by the water-based mud filtrate and can be used for the calculation of reservoir petrophysical parameters and the evaluation of pore structure. A long relaxation time represents large pores, and a short relaxation time represents small pores [18].

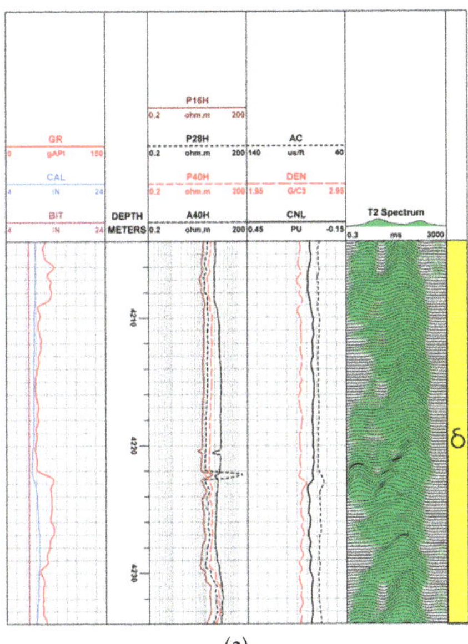 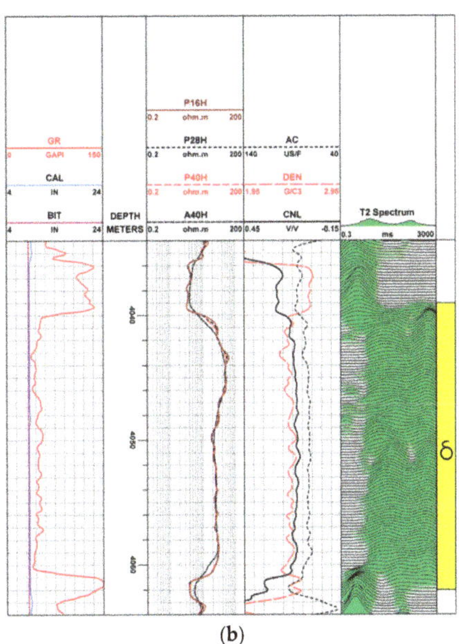

(a)   (b)

Figure 3. NMR logging response in different mud environments: (**a**) Water-based mud well A; (**b**) Oil-based mud well B.

Figure 3b is a comprehensive logging diagram of well B measured in the oil-based mud environment. The sandstone unit at interval 4039.0–4062.0 m is also a gas reservoir, but there is an obvious difference in NMR response characteristics from adjacent well A. Due to the influence of oil-based mud filtrate, there is a serious "tailing" phenomenon in the nuclear magnetic resonance T_2 spectrum of this horizon. The T_2 spectrum demonstrated a bimodal distribution. The T_2 value of the second main peak is approximately 1000 ms, and the maximum T_2 value is greater than 3000 ms. According to Figure 2c, the volume model of gas-bearing reservoirs in oil-based mud wells and the theoretical distribution of the nuclear magnetic T_2 spectrum, the distribution form of the nuclear magnetic T_2 spectrum in oil-based mud wells are obviously affected by the oil-based mud filtrate, so it is impossible to directly calculate reservoir petrophysical parameters.

2.2. Classification of Pore Structure Types of Reservoir Rocks

For the low-porosity and low-permeability reservoirs in the East China Sea, there is significant heterogeneity in the pore structure and lithology, and rocks with different lithologies and pore structures are affected by the invasion of oil-based mud filtrate. Therefore, the shape correction of the nuclear magnetic T_2 spectrum under the condition of oil-base mud needs to be based on the difference in the pore structure of the rock, and the morphology correction should be conducted separately for different reservoir rock pore structure types.

Through the observation and analysis of cast thin sections in this area, the pore structure of gas-bearing reservoir rocks can be divided into four categories according to the degree of pore development and connectivity. Figure 4a–d are the cast thin sections of typical plunger cores of reservoir rocks with four types of pore structures (hereinafter referred to as type I–IV rocks). The total porosity and permeability are the physical property test results of plunger cores. Type I rock clastic particles are well sorted, dominated by medium sand, rock pores are relatively developed, and pore connectivity is good. The clastic particles of type II rock are well sorted, dominated by medium sand, and the rock pores are relatively developed, but the connectivity is general. The clastic particles of type III rock are moderately sorted, slightly dominated by medium sand, and the rock pores are poorly developed and unevenly distributed. Type IV rock clastic particles are moderately sorted, slightly dominated by fine sand, poor rock pore development, and poor connectivity.

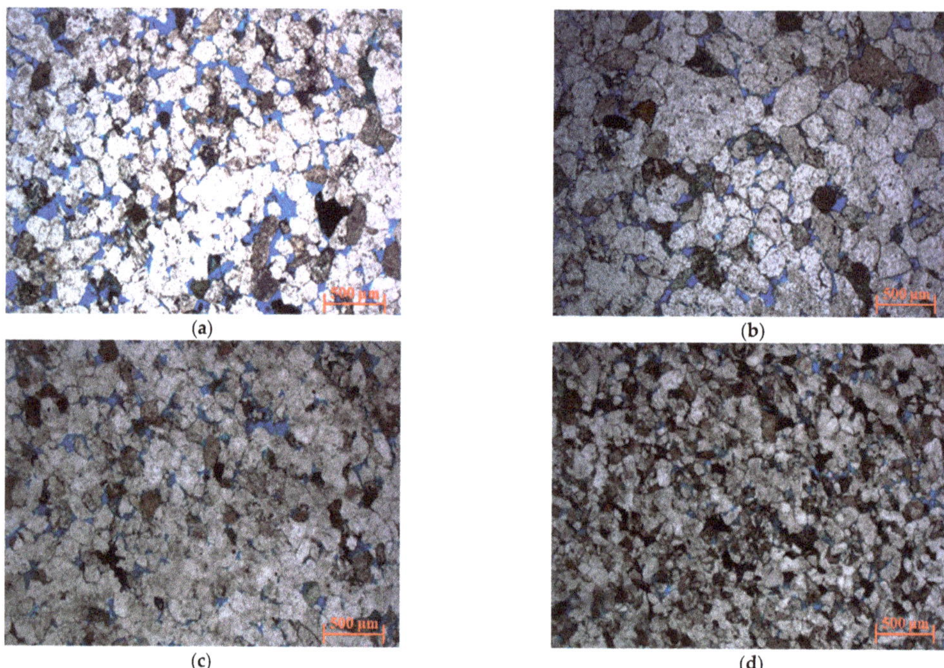

Figure 4. Casting sheet images of four types of rocks: (**a**) Type I (Φ = 13.9%, K = 269 mD); (**b**) Type II (Φ = 10.6%, K = 17.1 mD); (**c**) Type III (Φ = 10.4%, K = 4.9 mD); (**d**) Type IV (Φ = 9.7%, K = 0.45 mD).

Although the total porosity of cores decreases with the deterioration of pore connectivity, it is difficult to distinguish the types of reservoirs by porosity because the total porosity of cores of reservoirs with different pore structure types has little difference. Permeability reflects the seepage capacity of reservoir rocks and is a comprehensive characterization of porosity and pore structure [19]. Especially for reservoir rocks with low porosity and permeability, the permeability is primarily controlled by the pore structure in the rock, and the primary factor affecting the seepage capacity is the pores of different sizes and their matching relationship with the throat [20]. Figure 4 also shows that there are orders of magnitude differences in the permeability of reservoir cores with different pore structure types. Therefore, permeability can be used as the basis for classifying pore structure types. To facilitate the follow-up treatment, the pore structure types were divided according to the order of permeability: the permeability of type I rock is $K > 100$ mD, the permeability of type II rock is 10 mD $\leq K < 100$ mD, the permeability of type III rock is 1 mD $\leq K < 10$ mD, and the permeability of type IV rock is $K < 1$ mD. The results of the pore throat radius distribution calculated by the high-pressure mercury injection experiment can verify the feasibility of the pore structure classification standard. Figure 5 shows the pore throat radius distribution of 36 sandstones of four types of rocks divided according to the order of permeability, with obvious differences in pore throat dimensions. The main peak mean values of pore throat radius of type I–IV rocks are 16.13 µm, 4.74 µm, 1.35 µm and 0.55 µm, respectively. Although the main peak and mean values of the pore throat radius of a few cores are quite different, the overall pore throat radius distribution corresponds well with the four types of rocks. In addition, the pore throat radius distribution calculated by the high-pressure mercury injection experiment also explains the difference between the apparent porosity and total porosity of cast thin sections. For type IV rocks, micropores are relatively developed.

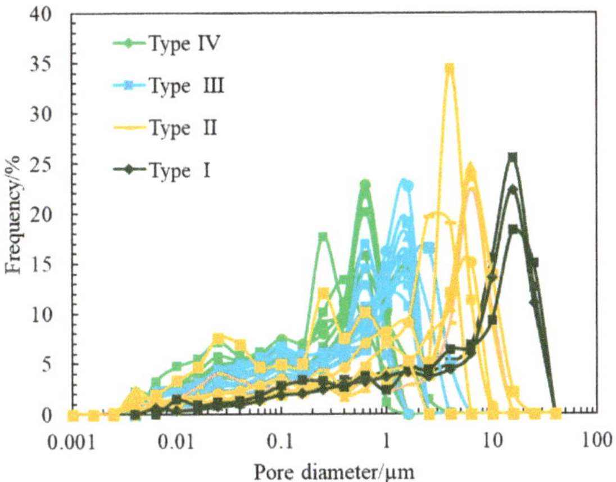

Figure 5. Pore-throat size distributions of four types of rocks.

The permeability affects the depth of oil-based mud filtrate invading the formation [21], and the pore fluid of rocks in different permeability formations has their own unique distribution characteristics [22]. Therefore, the T_2 spectrum of NMR logging will be significantly different for reservoir rocks with different types of pore structures under the condition of oil-based mud. To compare the difference in T_2 spectrum characteristic parameters between oil-based mud wells and water-based mud wells under similar formation physical parameters, this study selected the typical T_2 spectrum of nuclear magnetic resonance logging of four types of rocks (Figure 6), and the analysis of characteristic parameters is shown in Table 1. For the T_2 spectrum of water-based mud wells, the proportion of spectral area greater than 600 ms is very low. Therefore, the invasion degree of the oil-based mud filtrate can be reflected by counting the proportion of porosity with transverse relaxation times greater than 600 ms. The T_2 spectrum of nuclear magnetic resonance logging of type I rocks under different mud conditions had the greatest difference, indicating that for cores with good pore connectivity, the mud filtrate had the strongest change to the original formation fluid. Under the conditions of oil-based mud and water-based mud, the difference in T_2 spectrum of type II–IV rocks gradually decreased, indicating that with the deterioration of pore structure, the influence of mud filtrate on the fluid of the original formation gradually decreases, yet the shape of the T_2 spectrum still needed to be corrected.

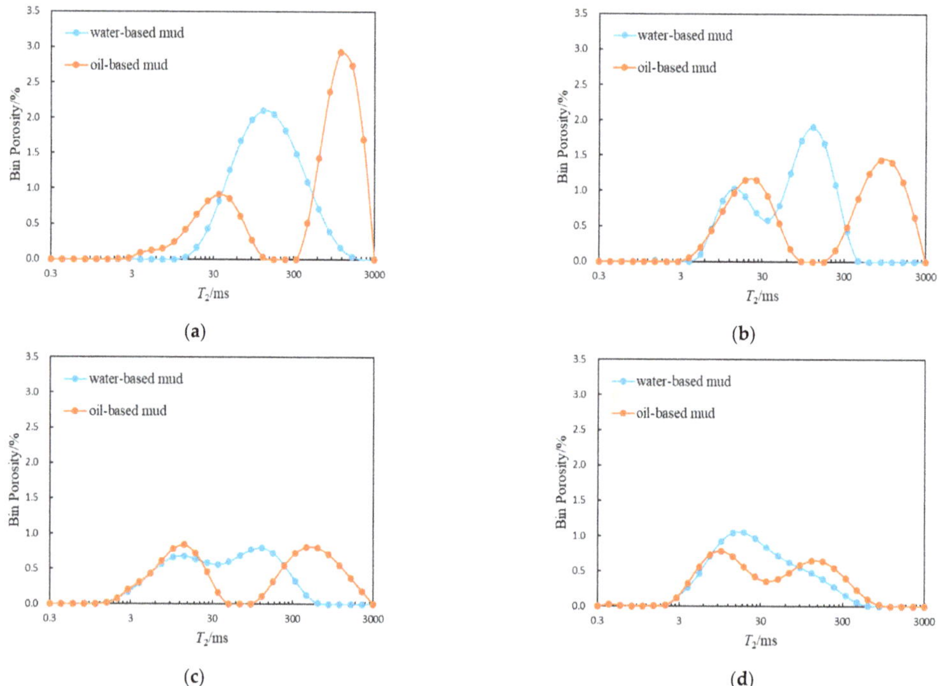

Figure 6. Comparisons of T_2 spectrum of four types of rocks in different mud environments: (**a**) Type I; (**b**) Type II; (**c**) Type III; (**d**) Type IV.

Table 1. T_2 spectrum characteristic parameters of four types of rocks in different mud environments.

Rock Structure Type	T_2 Peak/ms	T_2 Geometric Average/ms	Proportion of Pores Greater than 600 ms/%
I (Water-based mud)	125.26	140.97	8.07
I (Oil-based mud)	1156.99	363.55	66.12
II (Water-based mud)	125.26	58.92	0
II (Oil-based mud)	842.16	148.35	42.61
III (Water-based mud)	125.26	36.91	0.17
III (Oil-based mud)	446.21	82.47	26.85
IV (Water-based mud)	13.56	25.98	0.11
IV (Oil-based mud)	125.26	35.17	1.48

2.3. Determination of Optimal $T_{2cut\text{-}off}$

In this study, the core NMR experiment of water saturation and centrifugation was conducted on 29 cores to determine the optimal $T_{2cut\text{-}off}$ value in the study area. The specific parameters are shown in Table 2. According to the cut-off value, the T_2 spectrum of NMR logging was divided into macropores and small pores. The T_2 spectrum of NMR logging less than the T_2 cut-off value represents the small pores and retains its original shape; the T_2 spectrum of NMR logging greater than the T_2 cut-off value represents the macropore part, which needs morphological correction. The size of the $T_{2cut\text{-}off}$ value was related to the density of the core. With increasing porosity and permeability, the $T_{2cut\text{-}off}$ value also tended to increase, yet the fitting degree was relatively low (determination coefficient R2 = 0.38). Although the $T_{2cut\text{-}off}$ value had a good correlation with the geometric mean value of the T_2 spectrum (determination coefficient R2 = 0.68), the geometric mean value of the T_2 spectrum of the oil-based mud well was quite different from the geometric mean value of the T_2 spectrum of saturated water, so it cannot be applied to actual logging interpretation. Therefore, the average value of the experimental $T_{2cut\text{-}off}$ value of 29 cores of 17.48 ms was finally selected as the optimal $T_{2cut\text{-}off}$ value.

Table 2. Statistical table of physical property parameters of 29 sandstone core samples.

Core No	$T_{2cutoff}$/ms	Permeability/mD	NMR Porosity/%	T_2 Geometric Average/ms	Irreducible Water Saturation/%
1	10.35	10.80	17.09	11.91	38.74
2	15.70	15.30	16.73	20.96	34.25
3	13.66	17.51	16.57	16.06	44.48
4	44.48	522.3	21.10	101.65	23.12
5	23.81	63.10	18.4	35.50	31.74
6	12.75	5.40	12.35	6.63	55.37
7	22.21	18.45	17.26	22.97	55.37
8	12.75	3.20	12.88	9.48	53.47
9	5.54	9.90	14.9	8.50	43.86
10	15.70	4.20	14.38	12.39	51.74
11	18.04	4.20	8.78	2.36	77.81
12	14.64	4.83	15.11	8.32	57.03
13	12.75	2.18	14.22	9.98	54.19
14	22.21	0.43	9.82	16.38	44.48
15	19.33	0.20	8.66	5.57	73.46
16	5.54	6.56	14.54	5.51	53.65
17	12.75	1.73	13.39	10.08	51.82
18	16.83	1.23	12.13	9.91	55.3
19	38.72	68.40	17.58	37.19	46.32
20	18.04	2.87	15.47	14.73	49.42
21	4.82	1.65	13.32	6.40	47.75
22	5.94	39.40	14.11	13.88	38.10
23	11.09	2.50	13.65	9.40	51.29
24	10.35	18.20	15.64	15.37	42.69
25	6.36	3.25	13.3	8.97	44.96
26	3.65	2.85	11.09	6.33	43.44
27	27.36	29.70	18.31	29.38	39.56
28	47.68	1246.0	22.60	158.28	18.58
29	33.70	145.9	18.51	69.84	26.81

2.4. Case Studies

According to the morphological correction method of the T_2 spectrum of nuclear magnetic resonance logging under the condition of oil-based mud, the T_2 spectrum data of nuclear magnetic resonance logging under the condition of oil-based mud and water-based mud extracted by the multi-dimensional matrix method were used to calibrate four types of rocks, and the coefficients and constants of multivariate linear function corresponding to each type of rock were obtained. The executable program was compiled based on the FORTRAN language to correct the T_2 spectrum of NMR logging of oil-based mud wells (well B and well C) in the East China Sea. The results are shown in Figures 7 and 8.

According to the core porosity and permeability results, Figure 7 shows that the pore structure type of the well B reservoir in the study area is type III, belonging to a low permeability reservoir. It can be seen from the fifth track that the measured T_2 spectrum (Initial T2 spectrum) shows a continuous bimodal shape, representing that the maximum relaxation time of the second peak of the oil-based mud filtrate signal reaches 3000 ms, showing the characteristics of the macropore structure, which is seriously inconsistent with the core analysis results. Compared with the nuclear magnetic resonance T_2 spectrum before correction, the macropore part of the corrected nuclear magnetic T_2 spectrum (Corrected T2 spectrum) moves to the left, showing a continuous single peak shape, and the "tailing" phenomenon disappears, eliminating the fake long transverse relaxation time signal caused by the invasion of the oil-based mud filtrate. In addition, under the condition of oil-based mud, the permeability calculated by the T_2 spectrum of nuclear magnetic resonance logging before correction (Initial SDR permeability) is obviously larger than that of core analysis (Core-permeability), and the permeability calculated by T_2 spectrum of nuclear magnetic resonance logging after correction (Corrected SDR permeability) is smaller, which is in good agreement with the permeability results of core analysis.

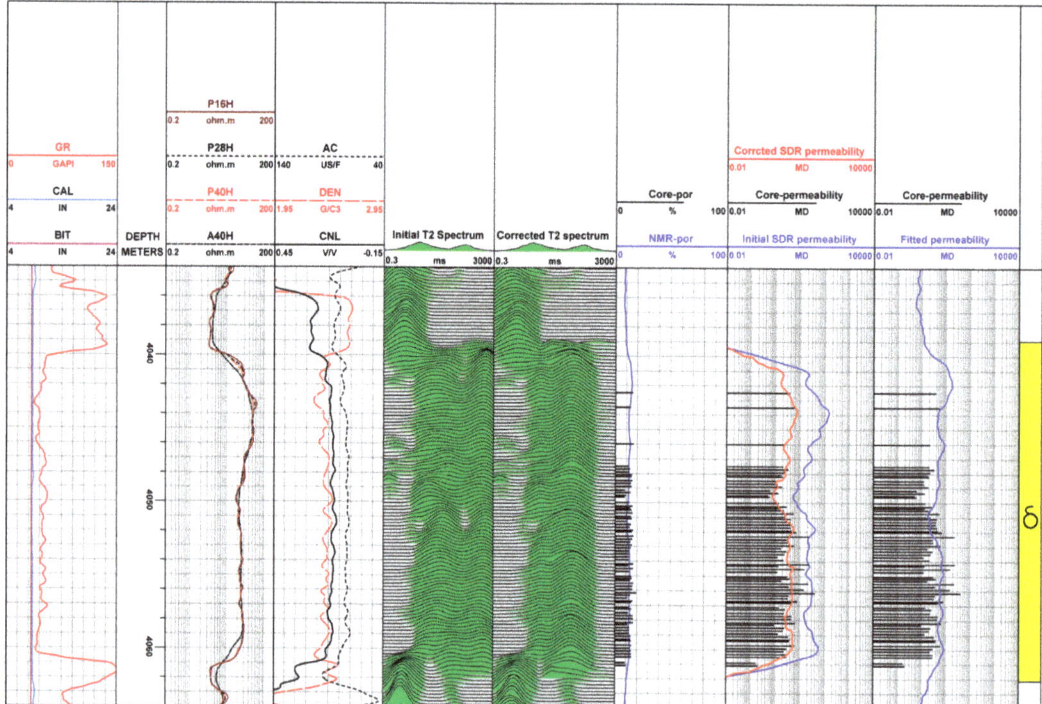

Figure 7. A field example of correcting the invasion of oil-based mud to field NMR logging in well B. In the first track, the displayed curves are gamma-ray (GR), borehole diameter (CAL), and bit diameter (BIT). The second track is the depth. The third is resistivity curve. The acoustic transit time log (AC), the density log (DEN), compensated neutron log (CNL) are shown in the fourth track. The fifth track is the initial measured T_2 spectrum of NMR logging before correction (Initial T_2 spectrum). The T_2 spectrum of nuclear magnetic resonance logging (Corrected T_2 spectrum) under the condition of water-based mud corrected by the method proposed in this study is displayed in the sixth track. The NMR calculated porosity (NMR-por) and core analysis porosity (Core-por) are shown in the seventh track and have good consistency. The Corrected SDR permeability is calculated permeability from the corrected T_2 spectrum by using Schlumberger-doll-Research (SDR) model, and the Initial SDR permeability is estimated permeability from the Initial T_2 spectrum by using the SDR model, the Core-permeability is derived permeability from core analysis. The ninth track is the permeability calculated by using the traditional porosity-permeability relationship (Fitted permeability). Good consistency of estimated permeability from the Corrected T_2 spectrum with core derived permeability illustrates the reliability of the proposed method.

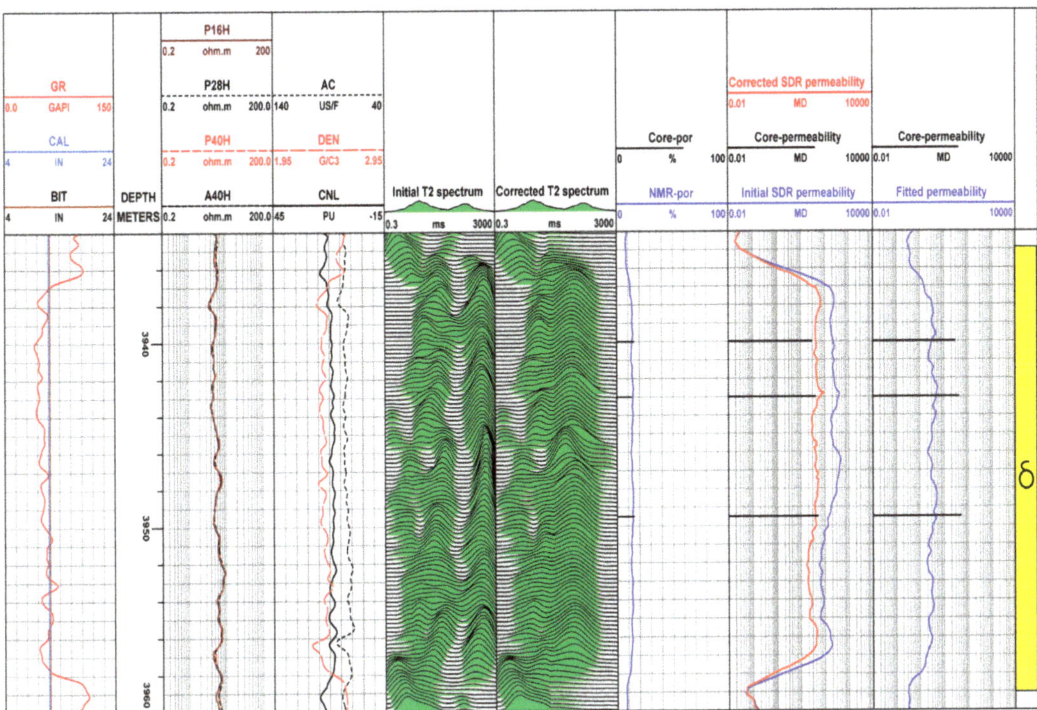

Figure 8. A field example of correcting the invasion of oil-based mud to field NMR logging in well C. In the first track, the displayed curves are gamma ray (GR), borehole diameter (CAL), and bit diameter (BIT). The second track is the depth. The third is resistivity curve. The acoustic transit time log (AC), the density log (DEN), compensated neutron log (CNL) are shown in the fourth track. The fifth track is the initial measured T_2 spectrum of NMR logging before correction (Initial T_2 spectrum). The T_2 spectrum of nuclear magnetic resonance logging (Corrected T_2 spectrum) under the condition of water based mud corrected by the method proposed in this study is displayed in the sixth track. The NMR calculated porosity (NMR-por) and core analysis porosity (Core-por) are shown in the seventh track and have good consistency. The Corrected SDR permeability is calculated permeability from the corrected T_2 spectrum by using Schlumberger-doll-Research (SDR) model, and the Initial SDR permeability is estimated permeability from the Initial T_2 spectrum by using SDR model, the Core-permeability is derived permeability from core analysis. The ninth track is the permeability calculated by using the traditional porosity-permeability relationship (Fitted permeability). Good consistency of estimated permeability from Corrected T_2 spectrum with core derived permeability illustrates the reliability of the proposed method.

According to the core analysis results, Figure 8 shows that the pore structure of well C is better than that of adjacent well B, and the pore structure is type II. It can be seen from the results of the fifth track that due to the improvement of pore structure and the increase of the number of macropores, the phenomenon of oil-based mud filtrate invading the formation pores is more serious. The nuclear magnetic T_2 spectrum shows a double peak or three peak shape, and the last two peaks even produce fractures and empty white belts. Compared with the nuclear magnetic resonance T_2 spectrum before correction, the macropore part of the corrected nuclear magnetic T_2 spectrum moves to the left, which not only eliminates the macropore illusion caused by the invasion of oil-based mud filtrate but also eliminates the phenomenon of the blank zone in the middle, making the nuclear magnetic T_2 spectrum continuous. In addition, under the condition of oil-based mud, the permeability calculated by the T_2 spectrum of nuclear magnetic resonance logging after correction also decreases, which is in good agreement with the permeability results of core analysis.

To further illustrate the reliability of the T_2 spectrum morphology correction results, the permeability results calculated by the T_2 spectrum of NMR logging before and after correction were compared with the permeability of core samples in wells B and C (Figure 9). The permeability calculated by the uncorrected T_2 spectrum of nuclear magnetic resonance logging obviously deviates from the 45° line, and the error with the core analysis result is approximately one order of magnitude (relative error 779.37%). The permeability calculated by the T_2 spectrum of corrected NMR logging is more consistent with that of core samples (relative error 34.32%), which is better distributed on both sides of the 45° line. Compared with the permeability calculated by core porosity and permeability fitting (relative error 82.98%), the overall accuracy of permeability calculated by the corrected nuclear magnetic T_2 spectrum improved by 48.66%. The results show that the morphological correction method of the T_2 spectrum of NMR logging under the condition of oil-based mud is reliable, and the influence of oil-based mud invasion on the T_2 spectrum of NMR logging has been eliminated. The corrected NMR T_2 spectrum can be used as the NMR T_2 spectrum measured under the condition of water-based mud, which lays a foundation for subsequent processing and interpretation.

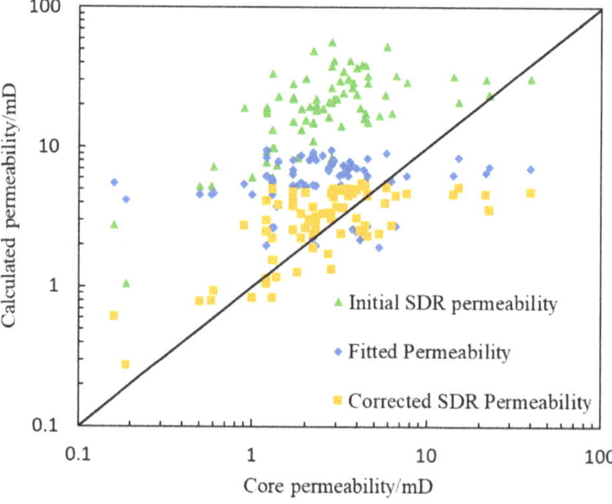

Figure 9. Comparison of permeability calculated by various methods and core analysis results.

3. Materials and Methods

The logging data used in this study were collected from five wells in the Xihu Sag, East China Sea basin, including three water-based mud wells and two oil-based mud wells (Table 3). The oil-based mud wells and water-based mud wells in the study area are measured by T_2 spectrum with the MREx nuclear magnetic resonance logging tool developed by Baker Hughes company, and the echo space is 0.6 ms. The maximum radial detection radius of the instrument is approximately 11.43 cm, which primarily detects the hydrogen signal of the fluid in the formation invasion zone. In addition, rock cast thin section image data and 29 rocks used for NMR experiments were obtained from these wells, and the data were provided by CNOOC Shanghai Branch.

Table 3. Summary of the wells logging information.

Well.	Mud Type	Lithology	Wellbore Orientation	Logs
A	WBM	Sand	Vertical	WL 3combo [1]+WLNMR
B	OBM	Sand	Vertical	LWD 3combo+WLNMR
C	OBM	Sand	Vertical	LWD 3combo+WLNMR
D	WBM	Sand	Vertical	WL 3combo+WLNMR
E	WBM	Sand	Vertical	WL 3combo+WLNMR

[1] 3 combo includes gamma-ray, density-neutron, resistivity.

3.1. Workflow of T_2 Spectrum Shape Correction Method

Since there are reliable technical means to evaluate reservoir parameters using nuclear magnetic resonance T_2 spectrum measured under water-based mud conditions, this study establishes the correlation between the T_2 spectrum of oil-based mud wells and the T_2 spectrum of water-based mud wells. The T_2 spectrum morphology of oil-based mud wells was corrected to the T_2 spectrum morphology of water-based mud wells. The realization method is shown in Figure 10. First, according to the difference in pore structure, the reservoir rocks were divided into four types. Second, the multidimensional matrix technology was used to construct the NMR T_2 spectrum sample library in the study area. Core saturation and centrifugal NMR experiments were used to determine the optimal T_2 cutoff value in the study area and then divide the large and small pores. Finally, under the condition of a similar pore structure, the multivariate linear function relationship of the T_2 spectrum of rock macropores in water-based mud wells and oil-based mud wells was established, and the correction of the rock T_2 spectrum morphology of oil-based mud wells was realized.

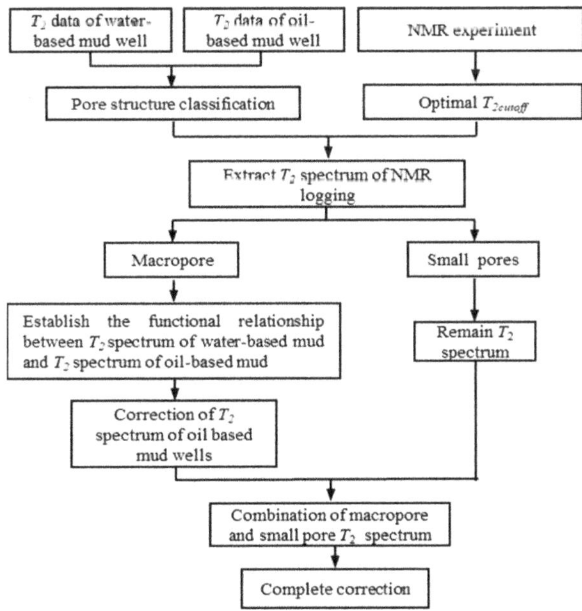

Figure 10. Workflow of the T_2 spectrum shape correction method for NMR logging under oil-based mud conditions.

3.2. Extraction of T_2 Spectrum from NMR Logging by the Multidimensional Matrix Method

To correct the morphology of the T_2 spectrum of formation nuclear magnetic resonance logging in an oil-based mud environment, it is necessary to determine the T_2 spectrum of nuclear magnetic resonance logging in a water-based mud environment, to establish the correlation between the T_2 spectrum of oil-based mud wells and the T_2 spectrum characteristics of water-based mud wells. However, for a well, nuclear magnetic resonance T_2 spectrum measurements in water-based mud and oil-based mud environments cannot be conducted at the same time. Additionally, due to the relatively small number of cores drilled during offshore operation, a large number of different types of mud invasion and nuclear magnetic resonance joint measurement experiments cannot be conducted. To obtain the nuclear magnetic T_2 spectrum in water-based mud and oil-based mud with similar pore structures, this study proposes the use of a multi-dimensional matrix method to construct a nuclear magnetic T_2 spectrum sample library in the study area to obtain T_2 spectrum data of NMR logging of reservoir rocks under the conditions of water-based mud and oil-based mud with similar curve values, such as porosity, permeability, natural gamma ray (GR), acoustic transit time log (AC), compensated neutron log (CNL), and density log (DEN). For the wells with two types of drilling fluid, the formation porosity is calculated by NMR logging. For water-based mud wells, because the T_2 spectrum of nuclear magnetic resonance logging can accurately calculate the permeability, the permeability calculated by the T_2 spectrum is used as the permeability of water-based mud wells; for oil-based mud wells, the permeability calculated by the T_2 spectrum is quite different from the core analysis results, but the difference between the permeability calculated by the porosity and permeability fitting method and the core analysis results is relatively small. Therefore, the permeability calculated by the porosity and permeability fitting method was used as the permeability of oil-based mud wells. The formula of the multidimensional matrix method is as follows:

$$A = [M(\phi, K, GR, AC, CNL, DEN), N(\phi, K, GR, AC, CNL, DEN)] \tag{2}$$

where, A is the total data matrix, including nuclear magnetic T_2 spectrum data in oil-based mud and water-based mud environments. M is the NMR T_2 spectrum sample library in an oil-based mud environment, and N is the NMR T_2 spectrum sample library in a water-based mud environment.

According to the different logging curve values of typical reservoirs in the study area, the nuclear magnetic T_2 spectrum data under the conditions of oil-based mud and water-based mud were classified into the corresponding sample library point by point, and the T_2 spectrum sample libraries under the conditions of water-based mud and oil-based mud are established. Therefore, for a certain depth point in the formation of oil-based mud wells, the T_2 spectrum of similar pore structure of adjacent water-based mud wells can be obtained, and then the characteristics of T_2 spectrum under different mud types can be compared to establish the correlation between them. This method avoids the problems of limited core NMR experimental data, poor representativeness, and a long experimental cycle, and is convenient for the subsequent calibration of the T_2 spectrum morphology correction model.

3.3. Development of T_2 Spectrum Shape Correction Model for Oil-Based Mud Wells

Both NMR T_2 spectrum data and core mercury injection data can be used to analyze the pore throat size of reservoir rocks, and the two have good consistency [23]. To quantitatively analyze pores of different sizes in rocks, some researchers have proposed a method to characterize the pore size by using the porosity of the T_2 spectrum interval of nuclear magnetic resonance logging [24]. Predecessors usually give seven fixed transverse relaxation time values to characterize the pore structure and pore size distribution information of reservoir rock, namely 1.0 ms, 3.0 ms, 10.0 ms, 33.0 ms, 100.0 ms, 300.0 ms, and 1000.0 ms, and the NMR logging T_2 spectrum is divided into eight porosity intervals (named 8 bins) [25]. Each interval reflects different pore sizes, the short transverse relaxation time represents small pores, and the long transverse relaxation time represents large pores.

Under the actual drilling differential pressure, macropores contribute the most to the permeability of the reservoir [26]. Mud filtrate primarily invades the macropores to displace the movable fluid (including movable water and oil and gas) within the detection range, while mud filtrate invades the small pores less, and the bound fluid will not change basically. Therefore, the invasion of mud filtrate only has a great impact on the T_2 spectrum of NMR logging corresponding to macropores but has little impact on the T_2 spectrum of NMR logging of bound fluid in small pores [27]. Therefore, this study corrected the movable fluid part of the T_2 spectrum of formation NMR logging after the invasion of oil-based mud filtrate and combined it with the T_2 spectrum of NMR logging of the original bound fluid part to obtain the complete NMR T_2 spectrum after morphological correction [28].

Considering that the invasion of oil-based mud had an impact on each pore component in the macropore space, and the average $T_{2cut\text{-}off}$ value is 17.48 ms, to characterize the impact of oil-based mud filtrate on the T_2 spectrum of NMR logging, this study defines only four transverse relaxation times: 33.0 ms, 100.0 ms, 300.0 ms and 1000.0 ms to divide the NMR T_2 spectrum. Combining the $T_{2cut\text{-}off}$ value and the maximum transverse relaxation time value, the T_2 spectrum can be divided into five intervals ([17.48, 33.0] ms, [33.0, 100.0] ms, [100.0, 300.0] ms, [300.0, 1000.0] ms, [1000.0, 3000.0] ms), and the pore composition percentage of each interval can be calculated, as shown in Equations (3)–(5) [29]:

$$X1 = \frac{\int_{T_{2cutoff}}^{T_{2(1)}} S(T)dt}{\int_{T_{2min}}^{T_{2max}} S(T)dt} \tag{3}$$

$$Xi = \frac{\int_{T_{2(i-1)}}^{T_{2(i)}} S(T)dt}{\int_{T_{2min}}^{T_{2max}} S(T)dt} \quad i = 2, 3, 4 \tag{4}$$

$$X5 = \frac{\int_{T_{2(4)}}^{T_{2max}} S(T)dt}{\int_{T_{2min}}^{T_{2max}} S(T)dt} \tag{5}$$

where, Xi is the percentage of pore components in the nuclear magnetic T_2 spectrum. The values of T_{2min} and T_{2max} are 0.3 ms and 3000 ms, respectively; $T_{2cutoff}$ is the optimal $T_{2cut\text{-}off}$ value obtained from the core NMR experiment, which is 17.48 ms in this study; $T_{2(i)}$ are the four T_2 relaxation time values defined above (33.0 ms, 100.0 ms, 300.0 ms and 1000.0 ms respectively); and $S(T)$ are the pore distribution functions of the nuclear magnetic T_2 spectrum.

Equations (3)–(5) can be used to calculate the percentages of five pore components according to the T_2 spectrum of NMR logging after the invasion of oil-based mud filtrate. The amplitude corresponding to each relaxation time of the measured NMR T_2 spectrum under the condition of water-based mud with T_2 relaxation time greater than 17.48 ms was defined as the dependent variable, and the percentage of five pore components was used as the independent variable. Therefore, the multivariate linear function relationship between the amplitude of each point of T_2 spectrum composition of NMR logging under the condition of water-based mud and the five pore components of T_2 spectrum of NMR logging under the condition of oil-based mud was established. Using this functional relationship, the amplitude of the nuclear magnetic resonance T_2 spectrum under water-based mud conditions with different relaxation times can be calculated from the nuclear magnetic resonance logging T_2 spectrum under oil-based mud conditions. The function relationship is shown in Equation (6):

$$A_1 = a_{11}X_1 + a_{12}X_2 + a_{13}X_3 + a_{14}X_4 + a_{15}X_5 + b_1$$
$$A_2 = a_{21}X_1 + a_{22}X_2 + a_{23}X_3 + a_{24}X_4 + a_{25}X_5 + b_2$$
$$A_3 = a_{31}X_1 + a_{32}X_2 + a_{33}X_3 + a_{34}X_4 + a_{35}X_5 + b_3 \tag{6}$$
$$\cdots$$
$$A_i = a_{i1}X_1 + a_{i2}X_2 + a_{i3}X_3 + a_{i4}X_4 + a_{i5}X_5 + b_i$$

where, A_i represents the amplitude value corresponding to the i th time distribution point of the T_2 spectrum of NMR logging after correction, and the value of i is the number of distribution points of the T_2 spectrum of NMR logging; $X_1, X_2 \ldots \ldots X_{15}$ is the percentage component of five pore components divided according to the corresponding reservoir type under the condition of oil-based mud; $a_1, a_2 \ldots \ldots a_{15}$ is the coefficient corresponding to the multivariate linear function corresponding to the i th distribution point, and its value is calibrated by T_2 spectrum data of nuclear magnetic resonance logging of water-based mud and oil-based mud in the sample library; $b_1, b_2 \ldots \ldots b_{15}$ is the constant corresponding to the multivariate linear function corresponding to the i th distribution point, and its value (Supplementary Materials, Tables S1–S4)is calibrated by T_2 spectrum data of nuclear magnetic resonance logging of water-based mud and oil-based mud in the sample library.

4. Conclusions

In the process of drilling, the invasion of oil-based mud filtrate has a serious impact on the T_2 spectrum of NMR logging, so NMR logging data cannot be directly used to evaluate reservoir parameters. Because the invasion degree of oil-based mud filtrate to reservoirs with different pore structures is different, through the analysis of cast thin sections, physical property experiments, and high-pressure mercury injection experiments, it is proposed to divide the reservoir rocks into four types based on the permeability range.

Based on the difference between the T_2 spectrum of nuclear magnetic resonance logging in water-based mud wells and oil-based mud wells, the corresponding morphological correction model of the nuclear magnetic resonance T_2 spectrum was established. By comparing the permeability, pore fitting permeability and core analysis permeability calculated by T_2 spectrum measured by NMR before and after correction, it was found that the permeability calculated after T_2 spectrum morphology correction is the most accurate, which improved the accuracy of calculating permeability by using NMR logging data in oil-based mud wells, it also laid a foundation for further NMR logging of data to evaluate other reservoir parameters.

The proposed correction model has been successfully applied in sandstone reservoirs, which lays a solid foundation for the morphological correction of the T_2 spectrum of NMR logging under the condition of oil-based mud in shale oil reservoirs. Different from sandstone reservoirs, when applying the model proposed in this study to NMR T_2 spectrum correction of shale oil reservoirs, it is necessary to comprehensively consider the complex pore network composition of shale oil reservoirs and the response mechanism of NMR logging under complex fluid occurrence modes.

Supplementary Materials: The following are available online. Tables S1–S4: The coefficients and constant matrix values used in the four types of rocks in the calibration process.

Author Contributions: Conceptualization, J.S. and J.C.; methodology, J.C. and J.L. (Jun Li); software, P.F.; validation, F.S., J.L. (Jing Lu); investigation, J.S.; resources, J.C.; data curation, P.F.; writing—original draft preparation, P.F. and W.Y.; writing—review and editing, J.S.; visualization, P.F.; supervision, J.L. (Jun Li). All authors have read and agreed to the published version of the manuscript.

Funding: This research was funded by National Natural Science Foundation of China, grant number 41874138 and 42004098, and the National Science and Technology Major Project, grant number 2016ZX05006002-004.

Data Availability Statement: Data available on request to the corresponding author.

Acknowledgments: The authors would like to thank Shanghai Branch of CNOOC Ltd. for the field logging data.

Conflicts of Interest: The authors declare no conflict of interest.

Sample Availability: Samples of the compounds are available from the authors.

References

1. Mohamed, A.; Salehi, S.; Ahmed, R. Significance and complications of drilling fluid rheology in geothermal drilling: A review. *Geothermics* **2021**, *93*, 102066. [CrossRef]
2. Adebayo, A.R.; Bageri, B.S.; Al Jaberi, J.; Salin, R.B. A calibration method for estimating mudcake thickness and porosity using NMR data. *J. Pet. Sci. Eng.* **2020**, *195*, 107582. [CrossRef]
3. Gamal, H.; Elkatatny, S.; Adebayo, A. Influence of mud filtrate on the pore system of different sandstone rocks. *J. Pet. Sci. Eng.* **2021**, *202*, 108595. [CrossRef]
4. Wu, J.; Fan, Y.; Wu, F.; Li, C. Combining large-sized model flow experiment and NMR measurement to investigate drilling induced formation damage in sandstone reservoir. *J. Pet. Sci. Eng.* **2019**, *176*, 85–96. [CrossRef]
5. Agwu, Q.E.; Akpabio, J.U. Using agro-waste materials as possible filter loss control agents in drilling muds: A review. *J. Pet. Sci. Eng.* **2018**, *163*, 185–198. [CrossRef]
6. Al-Arfaj, M.K.; Abdulraheem, A.; Sultan, A.; Amanullah; Hussein, I. Mitigating Shale Drilling Problems through Comprehensive Understanding of Shale Formations. In Proceedings of the Day 2 Mon, Doha, Qatar, 7 December 2015.
7. Elshehabi, T.; Ilkin, B. Well Integrity and Pressure Control in Unconventional Reservoirs: A Comparative Study of Marcellus and Utica Shales. In Proceedings of the SPE Eastern Regional Meeting, Canton, OH, USA, 13 September 2016.
8. Doak, J.; Kravits, M.; Spartz, M.; Quinn, P. Drilling Extended Laterals in the Marcellus Shale. In Proceedings of the Day 3 Tue, Pittsburgh, PA, USA, 9 October 2018; SPE: Richardson, TX, USA, 2018.
9. Gao, H.; Li, H. Determination of movable fluid percentage and movable fluid porosity in ultra-low permeability sandstone using nuclear magnetic resonance (NMR) technique. *J. Pet. Sci. Eng.* **2015**, *133*, 258–267. [CrossRef]
10. Mitchell, J.; Valori, A.; Fordham, E.J. A robust nuclear magnetic resonance workflow for quantitative determination of petrophysical properties from drill cuttings. *J. Pet. Sci. Eng.* **2019**, *174*, 351–361. [CrossRef]
11. Walsh, R.; Ramamoorthy, R.; Azhan, M. Applications of NMR Logging in Synthetic Oil-Base Muds in The Sirikit West Field, Onshore Thailand. In Proceedings of the SPWLA 42nd Annual Logging Symposium, Houston, TX, USA, 17 June 2001.
12. Zhao, Z.X.; Dong, C.M.; Zhang, X.G.; Lin, C.Y.; Huang, X.; Duan, D.P.; Lin, J.L.; Zeng, F.; Li, D. Reservoir controlling factors of the Paleogene Oligocene Huagang Formation in the north central part of the Xihu Depression, East China Sea Basin, China. *J. Pet. Sci. Eng.* **2019**, *175*, 159–172.
13. Hu, F.; Zhou, C.; Li, C.; Xu, H.; Zhou, F.; Si, Z. Water spectrum method of NMR logging for identifying fluids. *Pet. Explor. Dev.* **2016**, *43*, 268–276. [CrossRef]
14. Chen, J.; Hirasaki, G.; Flaum, M. NMR wettability indices: Effect of OBM on wettability and NMR responses. *J. Pet. Sci. Eng.* **2006**, *52*, 161–171. [CrossRef]
15. Chen, J.; Hirasaki, G.J.; Flaum, M. Effects of OBM Invasion on Irreducible Water Saturation: Mechanisms and Modifications of NMR Interpretation. In Proceedings of the SPE Annual Conference and Exhibition, Houston, TX, USA, 26 September 2004; 2004.
16. Marschall, D.M.; Coates, G. Laboratory MRI Investigation in the Effects of Invert Oil Muds on Primary MRI Log Determinations. In Proceedings of the SPE Annual Technical Conference and Exhibition, San Antonio, TX, USA, 5 October 1997.
17. Ighodalo, E.; Hursan, G.; McCrossan, J.; Belowi, A. Integrated NMR Fluid Characterization Guides Stimulation in Tight Sand Reservoirs. In Proceedings of the Day 3 Wed, Manama, Bahrain, 20 March 2019.
18. Al-Mahrooqi, S.; Grattoni, C.; Moss, A.; Jing, X. An investigation of the effect of wettability on NMR characteristics of sandstone rock and fluid systems. *J. Pet. Sci. Eng.* **2003**, *39*, 389–398. [CrossRef]
19. Schmitt, M.; Fernandes, C.P.; Wolf, F.G.; Neto, J.A.B.D.C.; Rahner, C.P.; dos Santos, V.S.S. Characterization of Brazilian tight gas sandstones relating permeability and Angstrom-to micron-scale pore structures. *J. Nat. Gas Sci. Eng.* **2015**, *27*, 785–807. [CrossRef]
20. Sakhaee-Pour, A.; Bryant, S.L. Effect of pore structure on the producibility of tight-gas sandstones. *AAPG Bull.* **2014**, *98*, 663–694. [CrossRef]
21. Fan, Y.; Wu, Z.; Wu, F.; Wu, J.; Wang, L. Simulation of mud invasion and analysis of resistivity profile in sandstone formation module. *Pet. Explor. Dev.* **2017**, *44*, 1045–1052. [CrossRef]
22. Zhao, J.; Yuan, S.; Li, W.; Ji, Y.; Liu, K. Numerical simulation and correction of electric logging under the condition of oil-based mud invasion. *J. Pet. Sci. Eng.* **2019**, *176*, 132–140. [CrossRef]
23. Xu, C.C.; Torres-Verodin, C. Quantifying fluid distribution and phase connectivity with a simple 3D cubic pore network model constrained by NMR and MICP data. *Comput. Geosci.* **2013**, *61*, 94–103. [CrossRef]

24. Han, Y.; Zhou, C.; Fan, Y.; Li, C.; Yuan, C.; Cong, Y. A new permeability calculation method using nuclear magnetic resonance logging based on pore sizes: A case study of bioclastic limestone reservoirs in the A oilfield of the Mid-East. *Pet. Explor. Dev.* **2018**, *45*, 183–192. [CrossRef]
25. Hursan, G.; Silva, A.; Van Steene, M.; Mutina, A. Learnings from a New Slim Hole LWD NMR Technology. In Proceedings of the Day 1 Mon, Abu Dhabi, United Arab Emirates, 9 November 2020; SPE: Richardson, TX, USA, 2020.
26. Xiao, Q.; Yang, Z.; Wang, Z.; Qi, Z.; Wang, X.; Xiong, S. A full-scale characterization method and application for pore-throat radius distribution in tight oil reservoirs. *J. Pet. Sci. Eng.* **2020**, *187*, 106857. [CrossRef]
27. Valle, B.; Bó, P.F.D.; Santos, J.; Aguiar, L.; Coelho, P.; Favoreto, J.; Arena, M.; dos Santos, H.N.; Ribeiro, C.; Borghi, L. A new method to improve the NMR log interpretation in drilling mud-invaded zones: A case study from the Brazilian Pre-salt. *J. Pet. Sci. Eng.* **2021**, *203*, 108692. [CrossRef]
28. Xiao, L.; Mao, Z.; Li, J.; Yu, H. Effect of hydrocarbon on evaluating formation pore structure using nuclear magnetic resonance (NMR) logging. *Fuel* **2018**, *216*, 199–207. [CrossRef]
29. Kenyon, W.E. Petrophysical principles of applications of NMR logging. *Log. Anal.* **1997**, *38*, 21–43.

molecules

Article

Multiphase Flow Regime Characterization and Liquid Flow Measurement Using Low-Field Magnetic Resonance Imaging

Rutger R. Tromp [1,2] and **Lucas M. C. Cerioni** [1,*]

1. KROHNE New Technologies, Kerkeplaat 12, 3313 LJ Dordrecht, The Netherlands; r.tromp@krohne.com
2. Department of Applied Physics, Eindhoven University of Technology, Den Dolech 2, 5600 MB Eindhoven, The Netherlands
* Correspondence: r.r.tromp@tue.nl

Abstract: Multiphase flow metering with operationally robust, low-cost real-time systems that provide accuracy across a broad range of produced volumes and fluid properties, is a requirement across a range of process industries, particularly those concerning petroleum. Especially the wide variety of multiphase flow profiles that can be encountered in the field provides challenges in terms of metering accuracy. Recently, low-field magnetic resonance (MR) measurement technology has been introduced as a feasible solution for the petroleum industry. In this work, we study two phase air-water horizontal flows using MR technology. We show that low-field MR technology applied to multiphase flow has the capability to measure the instantaneous liquid holdup and liquid flow velocity using a constant gradient low flip angle CPMG (LFA-CPMG) pulse sequence. LFA-CPMG allows representative sampling of the correlations between liquid holdup and liquid flow velocity, which allows multiphase flow profiles to be characterized. Flow measurements based on this method allow liquid flow rate determination with an accuracy that is independent of the multiphase flow profile observed in horizontal pipe flow for a wide dynamic range in terms of the average gas and liquid flow rates.

Keywords: low-field magnetic resonance; imaging; multiphase; flow measurement; pipe flow; two-phase flow; flow regime characterization; intermittent flow; slug flow; process and reaction monitoring

Citation: Tromp, R.R.; Cerioni, L.M.C. Multiphase Flow Regime Characterization and Liquid Flow Measurement Using Low-Field Magnetic Resonance Imaging. *Molecules* **2021**, *26*, 3349. https://doi.org/10.3390/molecules26113349

Academic Editor: Robert Brinson

Received: 30 April 2021
Accepted: 29 May 2021
Published: 2 June 2021

Publisher's Note: MDPI stays neutral with regard to jurisdictional claims in published maps and institutional affiliations.

1. Introduction

Within the petroleum industry there is a long-standing need for operationally robust, low cost, and real-time wellhead metering systems with accuracy across a broad range of produced volumes and hydrocarbon properties. Installations currently rely predominantly on accurate, yet costly, and operationally cumbersome test separators that by design deliver time-averaged multiphase flow rate data in well tests spanning several hours, thereby losing real-time flow information [1]. The real-time alternative to multiphase test separators, multiphase flow meter technology, has considerably improved in accuracy over the last decades. However, due to the complex combination of measurement technologies within these systems, these devises are highly sensitive to hydrocarbon properties and require repeated calibration in the field [2,3]. In addition, these multiphase flow meter systems tend to have a limited dynamic range in terms of produced volumes and associated multiphase flow profiles [1–3]. This poses problems in field applications as flow regimes can change over time due to natural production transients that can occur over the scale of hours or days, and inescapably occur over the lifetime of a well or due to flow restrictions caused by pipeline fouling that builds up over time during production [4]. In the absence of simple and accurate, plug-and-play well head metering solutions, many wells are operated with insufficient metering leading to suboptimal reservoir management and uncertainty in production allocation to individual wells [1–3].

For many years, magnetic resonance-based downhole logging tools have been successfully applied to in situ Earth formation evaluation [5]. These tools apply low field, time-domain magnetic resonance (MR) technology under challenging environmental conditions, proving the robustness of the technology. In the laboratory, the same technology can be used to obtain production fluid composition information from samples [6,7]. In the last decade, considerable effort has been spent to merge the two aspects and apply MR technologies in the process industry in pursuit of industry 4.0 compatible inline process monitoring and control [8]. Real-time wellhead metering systems are a concrete example of such inline process monitoring systems. Several MR technology-based research instruments have been developed for the petroleum industry, showing specific advantages of low-field MR technology when applied to multiphase flow measurement [1,9–12]. In this article, we show that a fully integrated multiphase flow meter [13] using low-field MR technology can act as a smart and robust measurement platform that has a large dynamic range in terms of produced volumes and associated multiphase flow profiles. This multiphase flow measurement platform can be applied as a general monitoring instrument in chemical and process control industries [8].

The complex flow profiles observed in multiphase flow emerge due to the differences in densities and viscosities of the fluid phases present in the flow [2]. For the case of two-phase, gas-liquid flows, these differences are maximum and the most challenging flow profiles occur. In this article, we focus on the horizontal pipe flow of water and air at atmospheric pressure. These fluids provide several advantages: Firstly, they are chemically safe, simplifying a flow loop design and operation; secondly, at atmospheric pressure the largest difference in gas and liquid density is obtained, leading to the most challenging flow profiles; and thirdly, water-based flows are relevant to a wide variety of processes in the chemical industry, including high water-fraction oil production in the petroleum industry.

In a static situation, there is a gravity induced separation between the two phases in a gas-liquid mixture present in a horizontal pipe: Gas is concentrated at the top of the pipe and liquid is concentrated at the bottom of the pipe, see Figure 1. When a pressure gradient is added along the length of the pipe, flow is induced. Since the phases have different densities and viscosities, the flow velocity associated with a given pressure drop per unit length is different for each phase. This so-called phase slip between phases is the primary complication in two-phase flows as it creates a dynamic pressure between the two phases. Depending on the cross-sectional area occupied by the two phases, which are commonly expressed using the dimensionless liquid holdup h_{liq}, see Figure 1, such that the cross-sectional area occupied by liquid is given by

$$A_{\mathrm{liq}} = h_{\mathrm{liq}} A_{\mathrm{pipe}}, \tag{1}$$

where $0 \leq h_{\mathrm{liq}} \leq 1$, and A_{pipe} is the pipe cross-sectional area, the surface tension of the fluid interface may or may not be strong enough to keep a stable interface between the two phases at a certain phase slip. If it is not, an instability in the local liquid holdup is induced. These instabilities can take the form of small, symmetric waves on a relatively stable fluid interphase, referred to as stratified flow; can give rise to large and chaotic wave patterns reminiscent of rough seas, referred to as wavy flow; and can even lead to such large instabilities that liquid is sucked up to the top of the pipe, creating so-called liquid slugs that are pushed along by the gas at high velocities.

Which flow pattern occurs in a given situation depends on many factors of which the upstream and downstream piping configuration are of paramount importance. For a given piping configuration and given volumetric liquid and gas flow rates, the flow pattern can be roughly estimated based on the superficial gas flow velocity, $u_{\mathrm{s,gas}} = Q_{\mathrm{gas}}/A_{\mathrm{pipe}}$, and the superficial liquid flow velocity, $u_{\mathrm{s,liq}} = Q_{\mathrm{liq}}/A_{\mathrm{pipe}}$, where Q_i is the volumetric flow rate of phase i. The superficial flow velocity thus represents the fictitious flow velocity of a single phase of a multiphase flow that it would have if all other phases in the multiphase flow were absent from the flow. Figure 2 shows an example flow map for two-phase, gas-liquid flow in a horizontal pipe section that uses the concept of the fictitious superficial flow velocity for parametrization [2]. The purple rectangle in Figure 2 indicates the superficial gas and liquid flow velocities that can be obtained using the multiphase flow loop used in this study. Details about this flow loop are presented in Section 4. Based on this flow map, we may expect stratified, wavy, and slug flow to be observed during multiphase flow experiments.

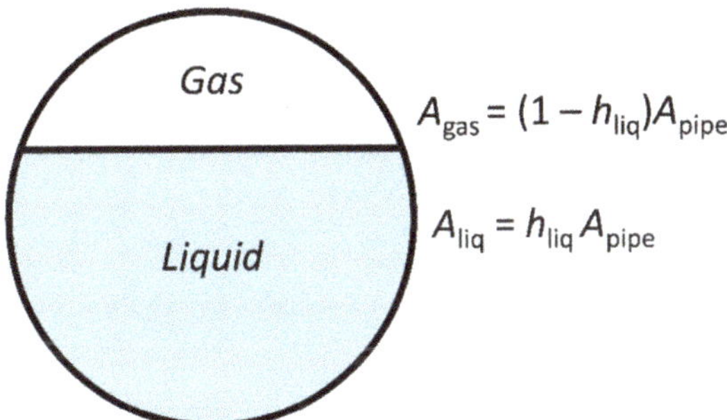

$$A_{gas} = (1 - h_{liq})A_{pipe}$$

$$A_{liq} = h_{liq} A_{pipe}$$

Figure 1. Schematic representation of the definition of the liquid holdup h_{liq}.

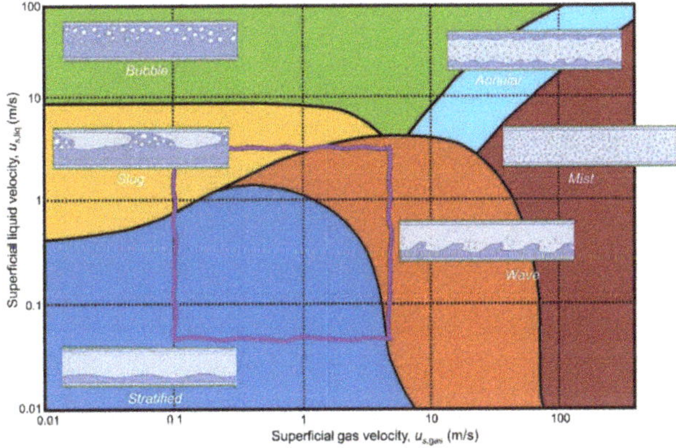

Figure 2. Schematic representation of a two-phase, gas-liquid flow map indicating the multiphase flow patterns likely to occur for a given combination of superficial flow velocities, adapted from [2]. The purple rectangle indicates the superficial gas and liquid flow velocities that can be obtained using the multiphase flow loop used in this study.

Based on the preceding discussion of multiphase flow patterns, one can see that correlations between instantaneous holdup and instantaneous flow velocity need to be characterized to accurately determine the flow rates of the individual phases in a two-phase flow.

An intuitive and simple method to measure the instantaneous liquid holdup would be to use the MR signal amplitude. However, for samples flowing through an industrial MR system this signal amplitude may depend on other factors than the liquid holdup alone. For instance, consider the CPMG pulse sequence [14,15] measurements presented in Figure 3 for a water-air slug flow-like flow regime. Both the amplitude at $t = 0$ s and the signal amplitude decay time of the CPMG signals vary considerably between measurements. Since air does not contribute to the CPMG signal in this experiment and water relaxation ($T_2 \sim 2–3$ s) is slow compared to the signal decay time, the signal decay time correlates with flow velocity u_{liq} [9]. The signal amplitude correlates strongly with h_{liq}, although the spin residence time in the polarizing magnetic field influences the observed signal amplitude as well, complicating the direct conversion of signal amplitudes to liquid holdups. The four CPMG signals highlighted in red in Figure 3 show that a given signal decay time or liquid flow velocity, can be observed for multiple signal amplitudes or liquid holdups. The major complication in multiphase flow measurement consequently is that the instantaneous flow rate, i.e.,

$$Q_{\text{liq}}(t) = u_{\text{liq}}(t)h_{\text{liq}}(t)A_{\text{pipe}}, \tag{2}$$

needs to be sampled in a way that ensures representative sampling of all characteristic flow events. When representative sampling is achieved, the average liquid flow rate <Q_{liq}> during a given time interval is given by the mean of the discrete set of flow events sampled during that time interval, i.e.,

$$Q_{\text{liq}} = u_{\text{liq}}(t)h_{\text{liq}}(t)A_{\text{pipe}} \cong u_{\text{liq},i}h_{\text{liq},i}A_{\text{pipe}}. \tag{3}$$

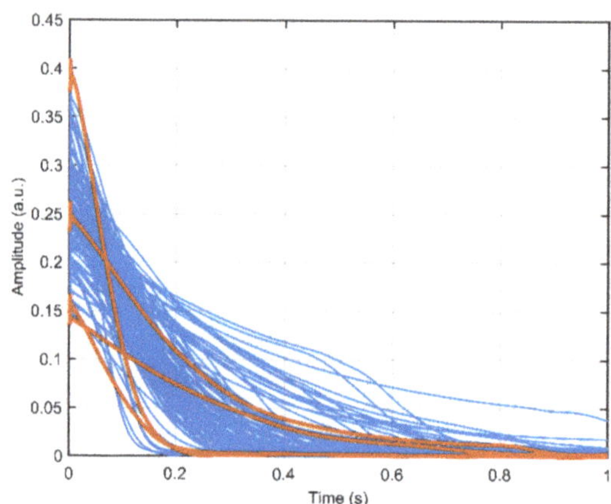

Figure 3. CPMG signal amplitude as a function of time shown for a set of measurements performed on a slug flow-like water-air multiphase flow. The two pairs of CPMG signals printed in red show that different signal amplitudes at *t* = 0 s can occur for the same signal amplitude decay time.

MR imaging (MRI) techniques may provide a direct measurement of liquid holdup, which is, for example, independent of the magnetic history of the sample. When MRI sequences are implemented, a spatially varying magnetic field or gradient **G**, is introduced in addition to the main magnetic field **B**$_0$. The effect of introducing the gradient **G** is that the resonance frequency of the nuclear spins varies with the position. The resonance frequency can thus be used to encode the position of nuclear spins. When MRI is applied in presence of flow, translational motion information can be extracted combining an imaging sequence with a spatially resolved measurement of molecular displacement. In many flow MRI studies, the velocity of a fluid media is measured by time-of-flight (TOF) [16–18] and phase shift methods [19]. A comprehensive review of non-medical flow MRI methods can be found in the articles by Gladden and Sederman [19,20]. The principles and relevant theory of flow MRI can be found in the books by Callaghan [21,22]. The fundamental concepts of MRI are discussed in an intuitive manner by McRobbie et al. [23].

Here, we focus on obtaining the bulk liquid flow velocity u_{liq} from the convective amplitude decay of the CPMG signals that is induced by the outflow of the excited sample volume [9]. We combine the CPMG pulse sequence with an external gradient G_z applied in the transversal vertical direction of the pipe to obtain a one-dimensional spatial distribution of the liquid, which we refer to as an one-dimensional (1D) distribution image. There are several techniques that combine the CPMG pulse sequence with an imaging sequence for spatial encoding [24]. As typical pulse sequences based on phase encoding gradients may increase the total acquisition time [24], we use frequency encoding to spatially encode all the points simultaneously during one CPMG spin-echo train. Since pulsed or modulated gradients require highly complex power electronics and gradient coils design that in an industrial application are translated into complexity for manufacturing, we use a constant-gradient CPMG [25]. This implementation additionally provides the advantage of short echo spacing for the convective amplitude decay velocity measurement. The frequency encoded spatially resolved 1D distribution image, can be obtained from the Fourier transform of each individual spin-echo signal [25].

To maximize the resolution and minimize the blurring effect due to inhomogeneities by spatial variations in $\mathbf{B}_0$ [26], we want to apply the maximum gradient strength available. During the application of the constant gradient G_z, the spectral width of the RF-pulses, Δv_{RF}, must be larger or equal to the spectral width of the sample, Δv_{sample}. The spectral width of the sample is given by $\Delta v_{sample} = \gamma G_z D / 2\pi$, where γ is the gyromagnetic ratio of the proton, and D is the pipe diameter [25]. The spectral width of a rectangular RF pulse can be approximated by $\Delta v_{RF} \approx 1/t_{pulse}$, where t_{pulse} is the RF pulse duration. The pulse sequence design relation between RF pulse length and applied gradient strength may thus be written as

$$t_{pulse} \leq \frac{2\pi}{\gamma G_z D}. \tag{4}$$

When we apply the maximum gradient strength G_z in our application, both 90° excitation and 180° refocusing pulses as used in a standard CPMG pulse sequence do not fulfill Equation (4). In other words, a standard CPMG would have limited bandwidth and cannot be used to excite and monitor the convective amplitude decay over the full pipe cross-section. This limitation was overcome by using a low flip angle CPMG (LFA-CPMG) [27], where all RF pulses are substituted by short duration pulses. This way the LFA-CPMG pulse sequence allows the instantaneous liquid holdup h_{liq} to be derived from the 1D liquid distribution image obtained from frequency encoded spin-echo signals, while the instant liquid flow velocity u_{liq} can be determined from the effective convective amplitude decay of the LFA-CPMG signals.

In this article, we will show that low-field MR technology applied to multiphase flow has the capability to measure the instantaneous liquid holdup and liquid flow velocity using the constant gradient LFA-CPMG pulse sequence. To this end, we applied the LFA-CPMG to study two-phase air-water flow experiments. The details of the experimental method and setup are presented in Section 4. In the following section it will be shown that LFA-CPMG allows the correlations between liquid holdup and liquid flow velocity to be determined, and it is shown that flow profiles can be identified based on these correlations. In addition, we show that flow calculations based on these correlations allow liquid flow rate determination with an accuracy that is independent of the multiphase flow profile observed in a horizontal pipe flow for a wide dynamic range in terms of the average gas and liquid flow rates.

2. Results

The set of two-phase air-water flow experiments that were performed is shown in Figure 4. Flow experiments have been performed for free flow and for flow disturbed by a downstream valve. This downstream ball valve closes in the vertical direction and was for 25% opened in the disturbed flow experiments. Flow regimes were identified for each flow experiment based on the multiphase flow profiles observed through a transparent pipe section. Stratified, wavy, and slug flow regimes were observed during the flow experiments and snapshots of typical gas and liquid phase distributions in these flows are indicated in Figure 2. Based on these visual identifications flow regime transition boundaries could be identified and these are indicated by the solid lines in Figure 4. Dashed lines indicate the gas volume fraction (GVF) of the multiphase flows, i.e., GVF = $Q_{gas}/(Q_{gas} + Q_{liq})$. Four experiments are highlighted by a black circle. These experiments are discussed in more detail in this article as examples. Video footage is made available in the supplementary information for some example experiments to illustrate dynamic liquid holdup variations occurring in gas-liquid multiphase flow.

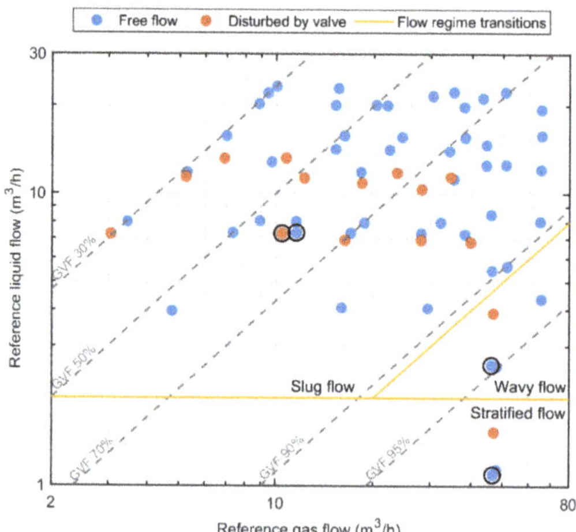

Figure 4. The measured single phase reference flow rates for all two-phase air-water flow experiments performed in this study indicated as dots. The dashed lines indicate the gas volume fraction (GVF) of the multiphase flows. Flow experiments have been performed for free flow and for flow disturbed by a downstream valve. Approximate flow regime transition boundaries were derived from visual inspection and are indicated by the solid lines. Four experiments that are discussed in more detail in this article are highlighted by a black circle.

In each flow, experiment data were acquired for 30 min using the low-field MR technology-based multiphase flow measurement method that is discussed in detail in Section 4. This measurement method uses a broadband excitation, constant-gradient LFA-CPMG pulse sequence to derive liquid holdup information from 1D liquid distribution images obtained from frequency encoded spin-echo signals, while liquid flow velocity information is derived from the convective amplitude decay of the LFA-CPMG signals with time as induced by the outflow of the excited sample volume. The frequency distribution of each spin-echo that is induced by the gradient field along the vertical direction represents a distribution image along the height of the pipe of the liquid portion of the flow, as air does not give an MR signal. This imaging functionality can be used to determine the multiphase flow profile in a given flow experiment from the combined liquid distribution images acquired during the 30 min of data acquisition.

2.1. Liquid Distribution Image Interpretation

Figure 5 shows the set of liquid distribution images acquired for the four experiments that are marked by a black circle in Figure 4. A surface representation of the liquid distribution images is used in which the images are sorted from the highest to lowest measured holdup to create a smooth surface that is more easily compared between experiments. For a full pipe of water, the liquid distribution image would take on the form of a semi-circle and these conditions occur for about 25% of the time in the slug flow experiment ($Q_{gas} = 11.5$ m³/h and $Q_{liq} = 7.3$ m³/h) shown in Figure 5a. The remainder of the time corresponds to a steady flow situation in which the pipe is partially filled with a constant liquid fraction. Slug flow can thus be envisioned as a binary flow system with two main events: Short bursts of liquid slugs with $h_{liq} \approx 1$, and longer events in which gas is accumulated at the top of the pipe and liquid at the bottom. This latter phase is very much comparable to the stratified flow experiment ($Q_{gas} = 46.6$ m³/h and $Q_{liq} = 1.1$ m³/h) presented in Figure 5d and is often referred to as the film phase of the slug flow.

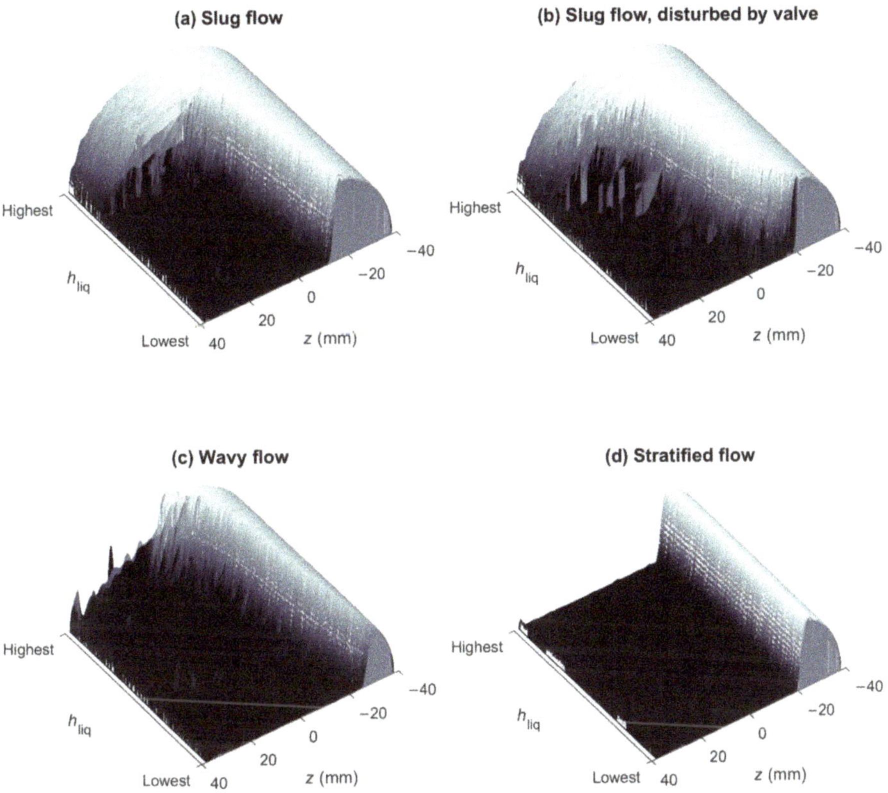

Figure 5. The set of liquid distribution images acquired in four different flow experiments represented as a surface plot in which liquid distribution images are sorted from the highest to lowest measured holdup. The axis labeled as z indicates the height along the flow tube with the pipe axis located at $z = 0$ mm. Each experiment corresponds to a unique multiphase flow profile: (**a**) Slug flow ($Q_{\text{gas}} = 11.5$ m^3/h and $Q_{\text{liq}} = 7.3$ m^3/h); (**b**) slug flow, disturbed by valve ($Q_{\text{gas}} = 10.4$ m^3/h and $Q_{\text{liq}} = 7.3$ m^3/h); (**c**) wavy flow ($Q_{\text{gas}} = 46.1$ m^3/h and $Q_{\text{liq}} = 2.6$ m^3/h); and (**d**) stratified flow ($Q_{\text{gas}} = 46.6$ m^3/h and $Q_{\text{liq}} = 1.1$ m^3/h).

The slug flow experiment was repeated with the flow disturbed by the downstream valve ($Q_{\text{gas}} = 10.4$ m^3/h and $Q_{\text{liq}} = 7.3$ m^3/h), in order to induce a more unstable flow profile. Figure 5b shows that although the flow profile can in general still be classified as slug flow, the valve disturbance leads to a considerably altered liquid distribution image surface, especially in the film phase. Although still about 15% of the time slugs with $h_{\text{liq}} \approx 1$ are observed, there is no longer a steady flow situation in the film phase. The film phase consequently has a liquid fraction in the pipe that changes continuously in time. This situation can be compared to the flow experiment labeled as wavy flow ($Q_{\text{gas}} = 46.1$ m^3/h and $Q_{\text{liq}} = 2.6$ m^3/h) that is presented in Figure 5c. The closing of the downstream valve in this flow experiment thus reduced the fraction of liquid slugs and induced wavy flow in the film phase of the slug flow.

2.2. Liquid Holdup and Velocity Correlations

As mentioned in the introduction, the accurate calculation of the liquid flow rate in multiphase flow comes down to the task of acquiring the correlations between the instantaneous liquid holdup and liquid flow velocity that are characteristic for a given flow profile. Figure 6 shows these correlations as derived from our low-field MR-based flow measurements for the same four flow experiments as for which the liquid distribution images were presented in Figure 5.

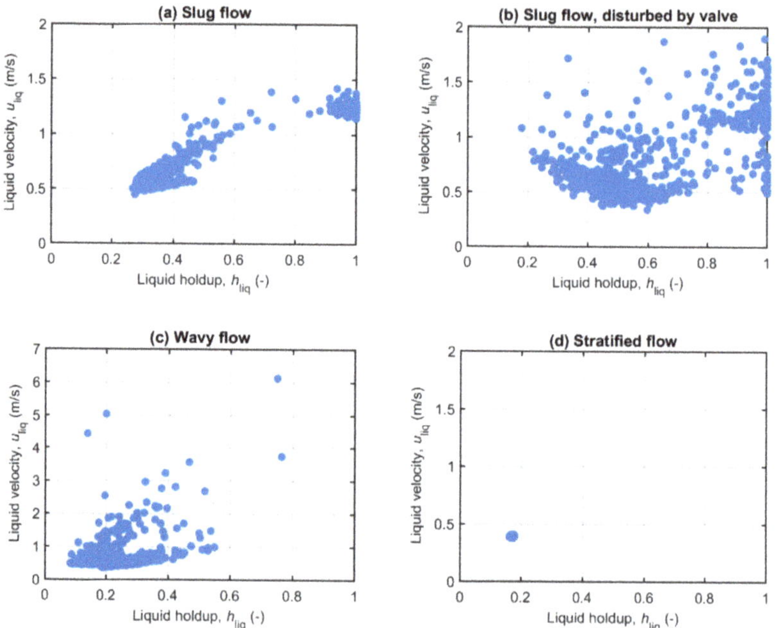

Figure 6. Measured liquid flow velocity as a function of measured liquid holdup for the same four flow experiments as for which the liquid distribution images were presented in Figure 5. These experiments correspond to: (**a**) Slug flow (Q_{gas} = 11.5 m^3/h and Q_{liq} = 7.3 m^3/h); (**b**) slug flow, disturbed by valve (Q_{gas} = 10.4 m^3/h and Q_{liq} = 7.3 m^3/h); (**c**) wavy flow (Q_{gas} = 46.1 m^3/h and Q_{liq} = 2.6 m^3/h); and (**d**) stratified flow (Q_{gas} = 46.6 m^3/h and Q_{liq} = 1.1 m^3/h).

Starting with the simplest case, stratified flow as presented in Figure 6d, a single point correlation is observed. This means that a given liquid holdup is directly related to a given liquid flow velocity. In such cases, the sampling rate and measurement time of the flow measurement method has little influence on the measurement results, as a single measurement already represents a representative sample of the multiphase flow. More structure is visible in the correlation plot for slug flow shown in Figure 6a. The binary character of slug flow is clearly represented by the two main concentrations of data points around h_{liq} = 0.3 (film phase) and around h_{liq} = 1 (slug phase). Note the higher flow velocity of about 2.5 times in the slug phase of the flow. Recalling that about 25% of the time the flow can be associated with the slug phase, most of the liquid flow is transported by the slug phase. This shows the importance of representative sampling of the flow, as even the minor under sampling of the slug phase may lead to large flow measurement errors.

The disturbed slug flow (Figure 6b) and wavy flow (Figure 6c) experiments exhibit a large spread in the flow velocities that are observed at a given liquid holdup. This spread consequently signals that complex stochastic processes are describing the correlations between the instantaneous liquid holdup and liquid flow velocity. Sufficiently fast sampling is expected to be very important for the accurate measurement of the liquid flow rate for these seemingly chaotic flow profiles. The fact that even for these flows clusters of data points are clearly observable in Figure 6, provides an indication that the statistics of these flows is sufficiently sampled, thus ensuring a representative sampling set of the liquid holdup and flow velocity correlations in the flow.

2.3. Liquid Flow Rate Measurement Accuracy

It is rather straightforward to compute the average liquid flow rate once the instantaneous liquid holdup and liquid flow velocity correlation is available. Assuming a statistically representative sample of the correlation is obtained by taking a total of N measurements, the average liquid flow rate may be computed as the average of the point-by-point product of liquid holdup and velocity, multiplied by the area of the pipe, i.e.,

$$\langle Q_{\text{liq}} \rangle = \frac{A_{\text{pipe}}}{N} \sum_{i=1}^{N} h_{\text{liq},i} u_{\text{liq},i}. \tag{5}$$

Figure 7 shows the relative liquid flow rate error as a function of the reference liquid flow rate for all experiments presented in Figure 4 together with an $\pm 5\%$ error band that is the generally accepted liquid flow rate accuracy required in multiphase flow metering (dashed lines) [12] and the relative error that corresponds to a zero-point inaccuracy of ± 1 m^3/h (dotted lines) that is commonly accepted as a practical limit for the accuracy of multiphase flow metering systems at low liquid flow rates [2]. The relative flow error is within the $\pm 5\%$ error band for all but four flow experiments. No difference in flow accuracy is observed between the flow experiments that had free flow or were disturbed by the downstream valve.

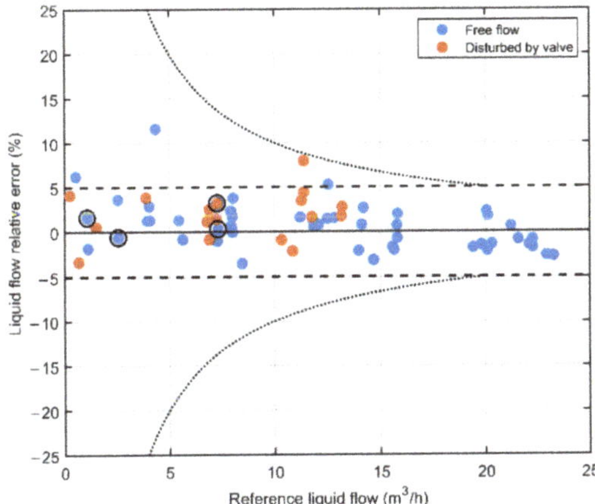

Figure 7. The relative liquid flow rate error as a function of reference liquid flow rate for all multiphase flow experiments that were presented in Figure 4. The dashed lines represent a $\pm 5\%$ error band that is the generally accepted liquid flow rate accuracy required in multiphase flow metering. The dotted lines represent the relative error that corresponds to a zero-point inaccuracy of ± 1 m^3/h that is commonly accepted as a practical limit for the accuracy of multiphase flow metering systems at low liquid flow rates.

3. Discussion

The multiphase flow profile independent liquid flow rate measurement accuracy presented in Section 2 is remarkable, considering the highly complex liquid holdup and liquid flow velocity correlations that are observed in these multiphase flows (see Figure 6), especially when the flow was disturbed by the downstream valve. Therefore, these results prove that a low-field MR-based flow metering apparatus can be applied to ensure representative sampling of the multiphase flow in a wide range of multiphase flow profiles and a wide dynamic range in terms of the average gas and liquid flow rates. In addition, this proves that flow regime identification is possible using MR measurement methods, which can be an important asset in the industry. For instance, in oil production and process optimization, where slugs may induce excessive structural vibration in piping systems causing component failures due to fatigue or resonance [2,28,29]. Research into the multiphase flow can benefit from measurement equipment that does not disturb the flow. Finally, the frequency encoding-based liquid holdup determination method applied in this study is shown to be robust enough to be applied to multiphase flows, opening up MR-based imaging opportunities in industrial multiphase flow applications.

4. Materials and Methods

The flow experiments presented in this work were performed using the M-PHASE 5000 multiphase flow meter developed by KROHNE [30] and shown in Figure 8. The 3.5 m long instrument is designed around a horizontal glass fiber reinforced epoxy (GRE) flow tube that is available in 2″, 3″, and 4″ pipe sizes. A 3″ pipe was used in this work, which has an 80 mm internal diameter. The main magnet section was constructed using a two-ring, 90 cm long, 0.2 T Halbach magnet with a length-to-radius ratio of 6. It contains a cylindrical region-of-interest (ROI) of 10 cm length and 10 cm diameter that was passively shimmed to a homogeneity of about 1000 ppm. A 12.5 cm long solenoid-shaped volume coil with an inner diameter of 12 cm, and a 40 cm long z-gradient coil with an inner diameter of 15 cm are centered on the ROI. The RF coil was used for both transmission of RF pulses and reception of NMR signals and was driven at 8.5 MHz using an RF power of 1.3 kW. The gradient coil was operated using a continuous current that generated a gradient field strength of 23.5 mT/m (equivalent to 10 kHz/cm). All electronics required for the NMR measurements and data transfer to a control computer are integrated into two flame-proof boxes that are mounted directly onto the flow meter. The instrument is additionally equipped with a pre-magnetization section consisting of 3 identical, two-ring, 30 cm long, unshimmed, 0.2 T Halbach magnets. The pre-magnetization length can be varied by selectively activating pre-magnetization sections by rotating the inner ring with respect to the outer ring in the Halbach section by 180 degrees.

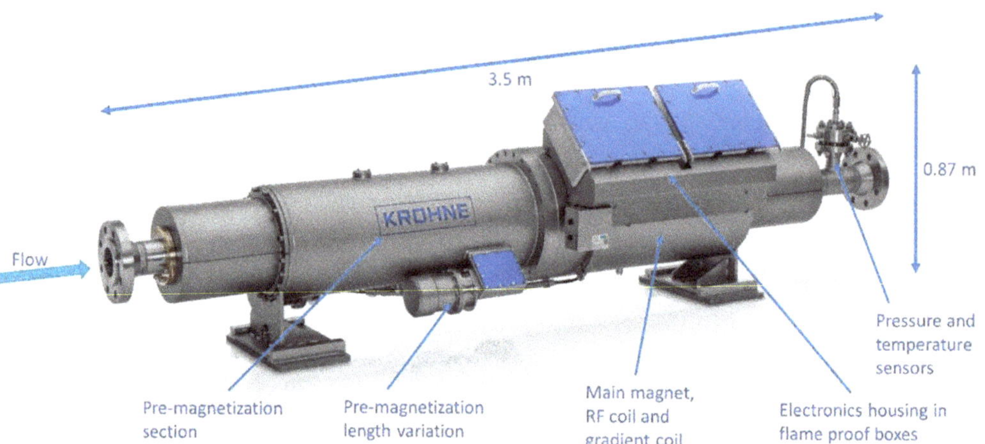

Figure 8. Annotated photograph of the KROHNE MPHASE 5000 MR-based multiphase flow meter used for the flow experiments presented in this work. Image courtesy of KROHNE [30].

The flow experiments presented in this work were performed on water-air mixtures for a wide range of flow rate combinations using the maximum pre-magnetization length. A schematic representation of the flow loop used is presented in Figure 9. Water flow was controlled by using 3 commercially available submersible garden water pumps that could be powered on independently. These pumps were placed in a 1 m³ Industrial Bulk Container (IBC) tank and yielded a combined maximum water flow rate of 48 m³/h. The flow loop was kept at atmospheric pressure via a vent in the IBC. The water injection point in the flow tubing for each pump was fitted with a ball valve that allowed for fine-tuning of the water injection for each individual pump. Whenever a pump was inactive, the ball valve allowed this pump to act as a controlled bypass for lowering flow rates through the magnet. This way, the superficial water flow velocity could be varied from about 0.5 cm/s up to 3 m/s. Air injection from a central laboratory compressed air supply was controlled using a needle valve and the superficial gas flow velocity could be varied from about 5 cm/s up to 3.5 m/s, corresponding to a maximum gas flow rate of 60 m^3/h through the multiphase flow meter. The reference volumetric flow rate for injected water was measured using a commercial electromagnetic flow meter (EMF in Figure 9) that has an accuracy better than 0.2% [31], while a commercial Coriolis flow meter with accuracy better than 0.5% [32] was used for air mass flow measurement. The air mass flow rate was converted to a volumetric air flow rate using dry air PVT calculations [33] based on the temperature and pressure measurements that are integrated into the M-PHASE 5000 multiphase flow meter. The flow loop piping layout was U-shaped and had a total straight flow length of 2 m (25D) applied upstream and downstream of the multiphase flow meter for flow conditioning. A ball valve was added just before the flow return connection to the IBC tank, which allowed the effect of flow disturbances on multiphase flow profiles and multiphase flow measurement accuracy to be studied by partially closing this valve. Multiphase flow profiles during the tests could be observed through the 1.5 m long transparent pipe section placed in front of the multiphase flow meter. Some example flow profile videos captures are provided in the supplementary information. Based on the observations during the tests, a flow map could be created to help predict the flow profile in the flow loop as a function of the gas and liquid flow rates. This flow map was presented as Figure 4 in Section 2.

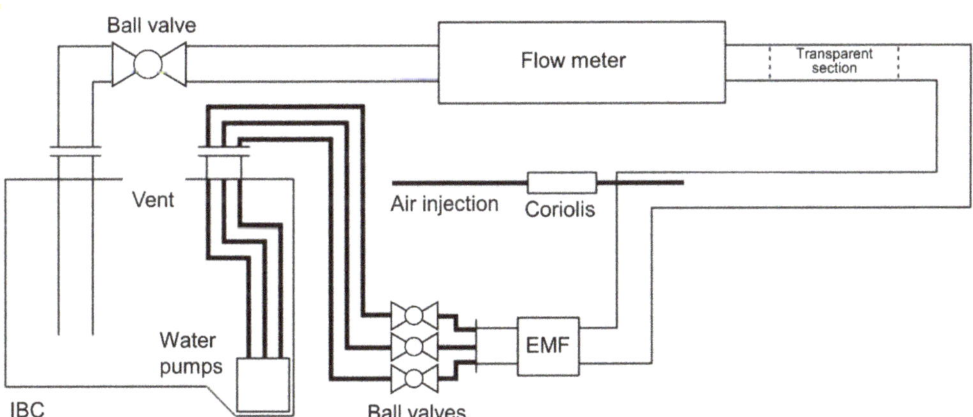

Figure 9. Schematic representation of the two-phase water-air flow loop used for the flow experiments presented in this work. Note that the air injection point, flow meter, and downstream ball valve of the flow loop were all placed at the same elevation above the IBC, making the piping horizontal over the entire length of the two-phase flow path.

Flow experiments were performed using broadband excitation constant-gradient LFA-CPMG pulse sequences [33,34] using 45° flip angle pulses of duration t_{pulse} = 10 μs, and 2τ = 800 μs echo spacing. This pulse sequence is shown schematically in Figure 10. To ensure the maximum initial signal amplitude and uniform spectral width of both excitation and refocusing pulses, pulse duration was kept the same for both excitation and refocusing pulses. Low flip angle pulse sequences can be used to determine the frequency spectrum of the sample in the ROI even in situations where limited SNR is available by combining the data from several echoes [35]. In addition, the amplitude decay of the LFA-CPMG signals with time due to the convective outflow of spins from the ROI, as obtained from the envelope of the spin-echo maximum amplitudes, can be used to derive average flow velocity information [9]. The number of acquired echoes and the wait time between consecutive pulse sequence executions were optimized in each flow experiment using the integrated flow measurement optimization feature of the KROHNE M-PHASE 5000. This algorithm actively tunes the number of echoes in real-time to match the lowest flow velocity component that occurs in the multiphase flow during the flow experiment. The wait time is set to 2 times the echo train length to ensure the sample is fully refreshed between consecutive pulse sequence executions. The liquid holdup was obtained by integrating the liquid distribution image obtained from the first 20 echoes and taking the ratio of this integral with the integral of a full pipe water reference measurement.

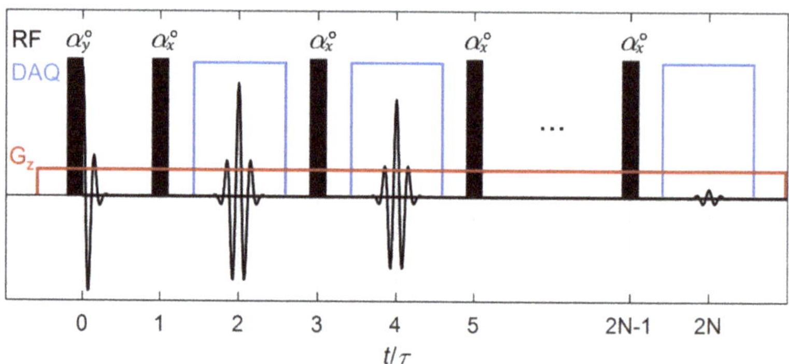

Figure 10. Schematic representation of the broadband excitation constant-gradient LFA-CPMG pulse sequence used in flow experiments. RF pulses with on-resonance flip angle α° are indicated by black rectangles. Digital acquisition (DAQ) of spin-echoes is represented in blue. The field gradient G_z is represented in dark red.

Prior to the two-phase flow experiments, velocity determination was calibrated, and liquid distribution image-based liquid holdup determination was validated.

Pure water flow experiments were performed on a dedicated single phase flow loop at KROHNE to calibrate the slope of the flow velocity determination via the convective decay of the LFA-CPMG signals. Figure 11 shows the relation between the reference flow velocity and the convective decay rate, R_v, for 17 different flow velocities up to 11 m/s. This convective decay rate was determined by fitting an exponential decay to $T_{2,\text{eff}}$-corrected LFA-CPMG signals. The $T_{2,\text{eff}}$ used for correction was determined as the effective T_2 decay obtained from a static LFA-CPMG experiment performed prior to each flow experiment. The use of an exponential convective decay model is based on the work by Petrova et al. [34] that showed the asymptotic form of the $T_{2,\text{eff}}$-corrected signal in low flip angle CPMGs to be exponential. The exponential fit was validated to be a better fit to our data than the linear fit method that is applied in flow measurement using bulk CPMGs.

The liquid holdup determination was validated by filling the multiphase flow loop shown in Figure 9 completely with water and by draining the piping in a controlled way via the ball valve that is placed downstream of the flow meter. A total of 19 different liquid levels were created this way, ranging from 100% down to 0.5% liquid holdup. The combined 1D liquid distribution images of these liquid level steps are shown in Figure 12, where the full semi-circle at the top left indicates the full pipe liquid experiment. When stepping from this 100% liquid holdup experiment down to lower liquid holdups, a progressively bigger portion of the semi-circle is cut-off due to the absence of liquid. As reference liquid holdup, the signal amplitude of a bulk spin-echo (A_{BSE}) was used. The reference liquid holdup for experiment i can be derived from the bulk spin-echo amplitude using the relation

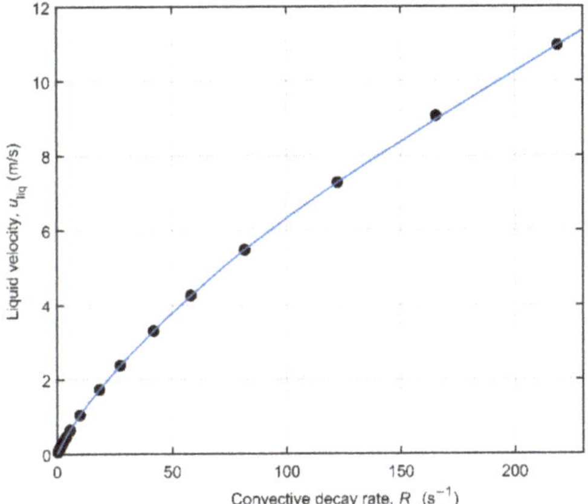

Figure 11. The reference liquid flow velocity, u_{liq}, as a function of the convective decay rate, R_v, that was obtained by fitting an exponential decay to $T_{2,eff}$-corrected LFA-CPMG signals. The solid line indicates the calibration function used in the two-phase flow experiments.

$$h_{liq,i} = \frac{A_{BSE,i}}{A_{BSE,100\%}},$$

(6)

in which $A_{BSE,100\%}$ is the bulk spin-echo amplitude obtained for a full pipe of liquid. The relation between the liquid holdup obtained using the bulk spin-echo and obtained from the liquid distribution images as acquired using the LFA-CPMG frequency-encoded spin-echoes is shown in Figure 13. A one-to-one correspondence between both methods exists over the entire range, indicating the robustness of the liquid distribution image-based liquid holdup determination method, even at liquid holdups down to a few percent.

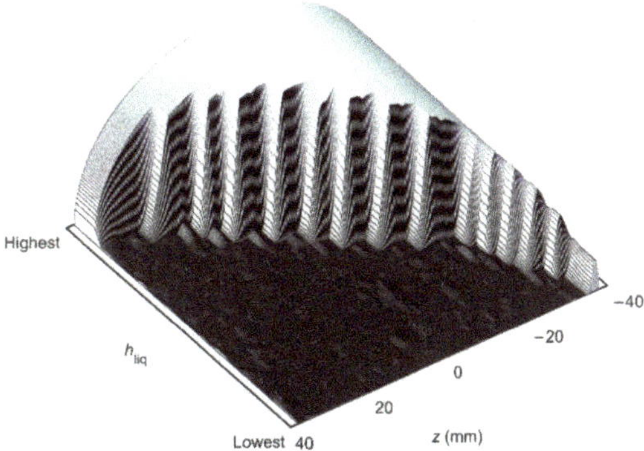

Figure 12. The set of 1D liquid distribution images acquired in the static verification experiments represented as a surface plot in which liquid distribution images are sorted from the highest to lowest measured holdup. The axis labeled as z indicates the height along the flow tube with the pipe axis located at $z = 0$ mm. The separate experiments are well recognized as steps in the surface as a progressively bigger portion of the semi-circle is cut-off due to the absence of liquid.

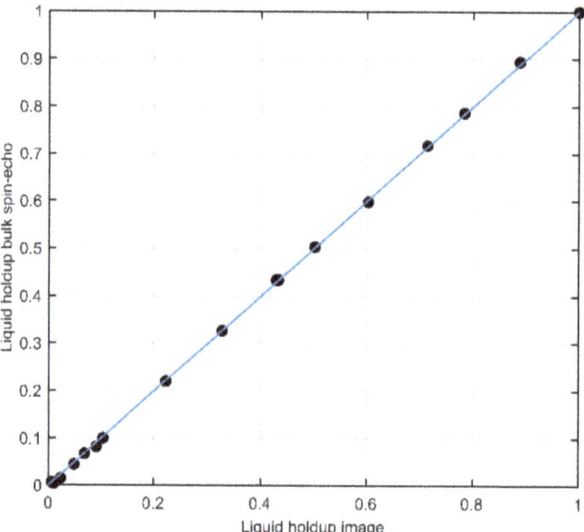

Figure 13. The liquid holdup as determined from a bulk spin-echo experiment as a function of the liquid holdup as determined from the 1D liquid distribution images obtained using a broad band constant-gradient LFA-CPMG pulse sequence. The measurements for 19 different liquid holdups show a 1-to-1 correspondence between the two methods.

5. Patents

Patent pending, provisional application number is DE 10 2021 111 162.5.

Supplementary Materials: The following videos of example flow experiments are available online. Video S1: Slug_slow_Qgas11.5_Qliq7.3.mp4 (Q_{gas} = 11.5 m^3/h and Q_{liq} = 7.3 m^3/h), Video S2: Slug_fast_Qgas43.2_Qliq21.2.mp4 (Q_{gas} = 43.2 m^3/h and Q_{liq} = 21.1 m^3/h), Video S3: Wavy_Qgas46.1 _Qliq2.6.mp4 (Q_{gas} = 46.1 m^3/h and Q_{liq} = 2.6 m^3/h), Video S4: Stratified_Qgas46.6_Qliq1.1.mp4 (Q_{gas} = 46.6 m^3/h and Q_{liq} = 1.1 m^3/h).

Author Contributions: Conceptualization, R.R.T. and L.M.C.C.; data curation, L.M.C.C.; methodology, R.R.T. and L.M.C.C.; software, L.M.C.C.; validation, R.R.T. and L.M.C.C.; formal analysis, R.R.T. and L.M.C.C.; investigation, R.R.T. and L.M.C.C.; visualization, R.R.T. and L.M.C.C.; writing—original draft, R.R.T.; writing—review and editing, R.R.T. and L.M.C.C. Both authors have read and agreed to the published version of the manuscript.

Funding: This research received no external funding.

Institutional Review Board Statement: Not applicable.

Informed Consent Statement: Not applicable.

Data Availability Statement: The data presented in this study are available on request from the corresponding author. The data are not publicly available due to KROHNE Messtechnik GmbH proprietary rights.

Acknowledgments: The authors would like to acknowledge Jankees Hogendoorn for his organizational support to the publication of this work and would like to thank the development team at KROHNE for their support in the development of the experimental facilities used for the flow measurements presented in this work. In this regard, the authors are especially indebted to Coert Kriger and Juan Pablo Nicoloff for their instrumental support during the experiments.

Conflicts of Interest: The authors declare no conflict of interest.

Sample Availability: Not available.

References

1. Appel, M.; Freeman, J.J.; Pusiol, D. Robust Multi-Phase Flow Measurement Using Magnetic Resonance Technology. In Proceedings of the SPE, Middle East Oil and Gas Show and Conference, Manama, Bahrain, 25–28 September 2011; SPE International: Richardson, TX, USA Paper Number: SPE-141465-MS. [CrossRef]
2. Dahl, E. *Handbook of Multiphase Flow Metering*, 2nd ed.; Norwegian Society for Oil and Gas Measurement: Oslo, Norway, 2005; ISBN 82-91341-89-3. Available online: nfogm.no/wp-content/uploads/2014/02/MPFM_Handbook_Revision2_2005_ISBN-82-9 1341-89-3.pdf (accessed on 30 April 2021).
3. Hansen, L.S.; Pedersen, S.; Durdevic, P. Multi-Phase Flow Metering in Offshore Oil and Gas Transportation Pipelines: Trends and Perspectives. *Sensors* **2019**, *19*, 2184. [CrossRef]
4. Makogon, T.Y. *Handbook of Multiphase Flow Assurance*, 1st ed.; Gulf Professional Publishing: Cambridge, MA, USA, 2019; ISBN 978-0-12-813062-9.
5. Kleinberg, R.L.; Jackson, J.A. An Introduction to the History of NMR Well Logging. *Concepts Magn. Reson.* **2001**, *13*, 340–342. [CrossRef]
6. Hirasaki, G.J.; Lo, S.-W.; Zhang, Y. NMR Properties of Petroleum Reservoir Fluids. *Magn. Reson. Imaging* **2003**, *21*, 269–277. [CrossRef]
7. Rudszuck, T.; Förster, E.; Nirschl, H.; Guthausen, G. Low-Field NMR for Quality Control on Oils. *Magn. Reson. Chem.* **2019**, *57*, 777–793. [CrossRef]
8. Meyer, K.; Kern, S.; Zientek, N.; Guthausen, G.; Maiwald, M. Process Control with Compact NMR. *Trends Anal. Chem.* **2016**, *83*, 39–52. [CrossRef]
9. Osán, T.M.; Ollé, J.M.; Carpinella, M.; Cerioni, L.M.C.; Pusiol, D.J.; Appel, M.; Freeman, J.; Espejo, L. Fast Measurements of Average Flow Velocity by Low-Field 1H NMR. *J. Magn. Reson.* **2011**, *209*, 116–122. [CrossRef] [PubMed]
10. Deng, F.; Xiao, L.; Liu, H.; An, T.; Wang, M.; Zhang, Z.; Wei, X.; Cheng, J.; Xie, Q.; Anferov, V. Effects and Corrections for Mobile NMR Measurement. *Appl. Magn. Reson.* **2013**, *44*, 1053–1065. [CrossRef]
11. Fridjonsson, E.O.; Stanwix, P.L.; Johns, M.L. Earth's Field NMR Flow Meter: Preliminary Quantitative Measurements. *J. Magn. Reson.* **2014**, *245*, 110–115. [CrossRef]
12. Zargar, M.; Johns, M.L.; Aljindan, J.M.; Noui-Mehidi, M.N.; O'Neill, K.T. Nuclear Magnetic Resonance Multiphase Flowmeters: Current Status and Future Prospects. *SPE Prod. Oper.* **2021**, 1–14. [CrossRef]
13. Bilgic, A.M.; Kunze, J.W.; Stegemann, V.; Hogendoorn, J.; Cerioni, L.; Zoeteweij, M. Multiphase Flow Metering with Nuclear Magnetic Resonance. *Tech. Messen.* **2015**, *82*, 539–548. [CrossRef]
14. Carr, H.Y.; Purcell, E.M. Effects of Diffusion on Free Precession in Nuclear Magnetic Resonance Experiments. *Phys. Rev.* **1954**, *94*, 630–638. [CrossRef]
15. Meiboom, S.; Gill, D. Modified Spin-Echo Method for Measuring Nuclear Relaxation Times. *Rev. Sci. Instrum.* **1958**, *29*, 688–691. [CrossRef]
16. Granwehr, J.; Harel, E.; Han, S.; Garcia, S.; Pines, A.; Sen, P.N.; Song, Y.-Q. Time-of-Flight Flow Imaging Using NMR Remote Detection. *Phys. Rev. Lett.* **2005**, *95*, 075503. [CrossRef]
17. Harel, E.; Granwehr, J.; Seeley, J.; Pines, A. Multiphase Imaging of Gas Flow in a Nanoporous Material Using Remote-Detection NMR. *Nat. Mater.* **2006**, *5*, 321–327. [CrossRef] [PubMed]
18. Ahmadi, S.; Mastikhin, I. Velocity Measurement of Fast Flows inside Small Structures with Tagged MRI. *Appl. Magn. Reason.* **2020**, *51*, 431–448. [CrossRef]
19. Gladden, L.F.; Sederman, A.J. Recent Advances in Flow MRI. *J. Magn. Reson.* **2013**, *229*, 2–11. [CrossRef] [PubMed]
20. Gladden, L.F.; Sederman, A.J. Magnetic Resonance Imaging and Velocity Mapping in Chemical Engineering Applications. *Annu. Rev. Chem. Biomol. Eng.* **2017**, *8*, 227–247. [CrossRef] [PubMed]
21. Callaghan, P.T. *Principles of Nuclear Magnetic Resonance Microscopy*; Oxford University Press: New York, NY, USA, 1993.
22. Callaghan, P.T. *Translational Dynamics and Magnetic Resonance*; Oxford University Press: New York, NY, USA, 2011.
23. McRobbie, D.W.; Moore, E.A.; Graves, M.J.; Prince, M.R. *MRI from Picture to Proton*, 3rd ed.; Cambridge University Press: Cambridge, UK, 2017.
24. Muir, C.E.; Balcom, B.J. A Comparison of Magnetic Resonance Imaging Methods for Fluid Content Imaging in Porous Media. *Magn. Reson. Chem.* **2013**, *51*, 321–327. [CrossRef]
25. Hertel, S.A.; De Kort, D.W.; Bush, I.; Sederman, A.J.; Gladden, L.F.; Anger, B.; De Jong, H.; Appel, M. Fast Spatially-Resolved T2 Measurements with Constant-Gradient CPMG. *Magn. Reson. Imaging* **2019**, *56*, 70–76. [CrossRef]
26. Brown, R.W.; Cheng, Y.C.N.; Haacke, E.M.; Thompson, M.R.; Venkatesan, R. *Magnetic Resonance Imaging: Physical Principles and Sequence Design*, 2nd ed.; John Wiley & Sons: Hoboken, NJ, USA, 2014; pp. 272–283.
27. de Andrade, F.D.; Netto, A.M.; Colnago, L.A. Qualitative Analysis by Online Nuclear Magnetic Resonance Using Carr-Purcell-Meiboom-Gill Sequence with Low Refocusing Flip Angles. *Talanta* **2011**, *84*, 84–88. [CrossRef]
28. Havre, K.; Stornes, K.O.; Stray, H. Taming Slug Flow in Pipelines. *ABB Rev.* **2000**, *4*, 55–63.
29. Coker, A.K. Fluid Flow. In *Ludwig's Applied Process Design for Chemical and Petrochemical Plants*, 4th ed.; Coker, A.K., Ed.; Gulf Professional Publishing-Elsevier: Oxford, UK, 2007; Volume 1, pp. 133–202.
30. KROHNE Messtechnik GmbH, Duisburg, Germany.
31. KROHNE Optiflux 3300 Electromagnetic Flow Meter.

32. KROHNE Optimass 6400 Coriolis Mass Flow Meter.
33. Picard, A.; Daves, R.S.; Gläser, M.; Fujii, K. Revised Formula for the Density of Moist Air (CIPM-2007). *Metrologica* **2008**, *45*, 149–155. [CrossRef]
34. Petrova, M.V.; Doktorov, A.B.; Lukzen, N.N. CPMG Echo Amplitudes with Arbitrary Refocusing Angle: Explicit Expressions, Asymptotic Behavior, Approximations. *J. Magn. Reson.* **2011**, *212*, 330–343. [CrossRef] [PubMed]
35. Lu, D.; Joseph, P.M. A Matched Filter Echo Summation Technique for MRI. *Magn. Reson. Imaging* **1995**, *13*, 241–249. [CrossRef]

molecules

Article

^{1}H NMR Study of the HCa$_2$Nb$_3$O$_{10}$ Photocatalyst with Different Hydration Levels

Marina G. Shelyapina [1,*], Oleg I. Silyukov [2], Elizaveta A. Andronova [1], Denis Y. Nefedov [1], Anastasiia O. Antonenko [1], Alexander Missyul [3], Sergei A. Kurnosenko [2] and Irina A. Zvereva [2]

[1] Faculty of Physics, Saint Petersburg State University, 7/9 Universitetskaya nab., 199034 Saint Petersburg, Russia; st064749@student.spbu.ru (E.A.A.); d.nefedov@spbu.ru (D.Y.N.); a.antonenko@spbu.ru (A.O.A.)

[2] Institute of Chemistry, Saint Petersburg State University, 7/9 Universitetskaya nab., 199034 Saint Petersburg, Russia; oleg.silyukov@spbu.ru (O.I.S.); st040572@student.spbu.ru (S.A.K.); irina.zvereva@spbu.ru (I.A.Z.)

[3] CELLS-ALBA Synchrotron, 08290 Cerdanyola del Vallès, Barcelona, Spain; amissiul@cells.es

* Correspondence: marina.shelyapina@spbu.ru

Abstract: The photocatalytic activity of layered perovskite-like oxides in water splitting reaction is dependent on the hydration level and species located in the interlayer slab: simple or complex cations as well as hydrogen-bonded or non-hydrogen-bonded H$_2$O. To study proton localization and dynamics in the HCa$_2$Nb$_3$O$_{10}\cdot y$H$_2$O photocatalyst with different hydration levels (hydrated—α-form, dehydrated—γ-form, and intermediate—β-form), complementary Nuclear Magnetic Resonance (NMR) techniques were applied. ^{1}H Magic Angle Spinning NMR evidences the presence of different proton containing species in the interlayer slab depending on the hydration level. For α-form, HCa$_2$Nb$_3$O$_{10}\cdot1.6$H$_2$O, ^{1}H MAS NMR spectra reveal H$_3$O$^+$. Its molecular motion parameters were determined from ^{1}H spin-lattice relaxation time in the rotating frame ($T_{1\rho}$) using the Kohlrausch-Williams-Watts (KWW) correlation function with stretching exponent $\beta = 0.28$: $E_a = 0.210(2)$ eV, $\tau_0 = 9.0(1) \times 10^{-12}$ s. For the β-form, HCa$_2$Nb$_3$O$_{10}\cdot0.8$H$_2$O, the only ^{1}H NMR line is the result of an exchange between lattice and non-hydrogen-bonded water protons. $T_{1\rho}(1/T)$ indicates the presence of two characteristic points (224 and 176 K), at which proton dynamics change. The γ-form, HCa$_2$Nb$_3$O$_{10}\cdot0.1$H$_2$O, contains bulk water and interlayer H$^+$ in regular sites. ^{1}H NMR spectra suggest two inequivalent cation positions. The parameters of the proton motion, found within the KWW model, are as follows: $E_a = 0.217(8)$ eV, $\tau_0 = 8.2(9) \times 10^{-10}$ s.

Keywords: layered perovskite-like niobate; Dion-Jacobson phase; proton NMR

check for updates

Citation: Shelyapina, M.G.; Silyukov, O.I.; Andronova, E.A.; Nefedov, D.Y.; Antonenko, A.O.; Missyul, A.; Kurnosenko, S.A.; Zvereva, I.A. ^{1}H NMR Study of the HCa$_2$Nb$_3$O$_{10}$ Photocatalyst with Different Hydration Levels. *Molecules* **2021**, *26*, 5943. https://doi.org/10.3390/molecules26195943

Academic Editor: Elena G. Bagryanskaya

Received: 30 August 2021
Accepted: 28 September 2021
Published: 30 September 2021

Publisher's Note: MDPI stays neutral with regard to jurisdictional claims in published maps and institutional affiliations.

1. Introduction

In recent years, layered perovskite-like oxides have attracted much attention because of their outstanding physical and chemical properties, including high-temperature superconductivity [1,2], colossal magnetoresistance [3], the capability of photocatalytic water decomposition under sunlight irradiation for further hydrogen storage [4,5], and ionic conductivity due to high mobility of interlayer cations [6,7]. The majority of ion-exchangeable layered perovskite-like oxides can be converted into their protonated forms, which, besides being proton conductors [6,8] and photocatalysts for water splitting [9–11], exhibit the ability to intercalate water [10–14] and other molecules [15,16] and/or to form graft derivatives [16–19] susceptible to further exfoliation [10,20,21].

The hydrated form of HCa$_2$Nb$_3$O$_{10}$ (usually referred to as HCa$_2$Nb$_3$O$_{10}\cdot1.5$H$_2$O in the literature) belongs to the Dion-Jacobson phase and can be obtained from KCa$_2$Nb$_3$O$_{10}$ oxide by ion-exchange in acid solutions [22]. It was shown that HCa$_2$Nb$_3$O$_{10}\cdot1.5$H$_2$O enables the intercalation of amines by an acid-base mechanism [23] and may be later exfoliated into nanolayers [24,25]. Both KCa$_2$Nb$_3$O$_{10}$ and HCa$_2$Nb$_3$O$_{10}\cdot1.5$H$_2$O, as well as their exfoliated and restacked forms, exhibit photocatalytic properties [26–28]. Along with this form, there may be others with a lower water content. The ability to intercalate water molecules often

39

plays a crucial role in other intercalation reactions and photocatalysis [9,29–31]. Hydrated protonated forms may comprise protons [13,30] or charged complexes like $H^+ \dots n \cdot H_2O$ in their interlayer slab [32–34]. Obviously, water content and its state and localization should affect both the pathway and efficiency of chemical or photocatalytic reactions. From this perspective, an identification of proton-containing species and a comprehensive study of their motion in the interlayer slab is required.

Proton Nuclear Magnetic Resonance (NMR) is one of the most versatile experimental methods. It enables the identification of the proton-containing species and provides insight on the local structure [13,32,35–37] and information at the microscopic level on the dynamics of intercalated species [13,16,32,35,36,38]. In particular, by using ^{1}H NMR, it was shown that both the local environment and the dynamics of hydrogen in these materials are affected by the stacking sequence of the perovskite-like slabs [39].

Here, we report on the results of the proton NMR spectroscopy and relaxation studies of the layered perovskite-like niobate $HCa_2Nb_3O_{10}$ with different hydration levels: hydrated—α-form, dehydrated—γ-form, and intermediate—β-form. The details of their synthesis can be found in Section 3.

2. Results and Discussion

2.1. X-ray and TG Analysis

Figure 1a shows the X-ray Diffraction (XRD) patterns of the studied $HCa_2Nb_3O_{10} \cdot yH_2O$ samples with different hydration levels. The XRD shows that the samples are practically monophase. All the samples can be described by the $P4/mmm$ space group. The unit cell is shown in Figure 1b. The lattice parameters are listed in Table 1.

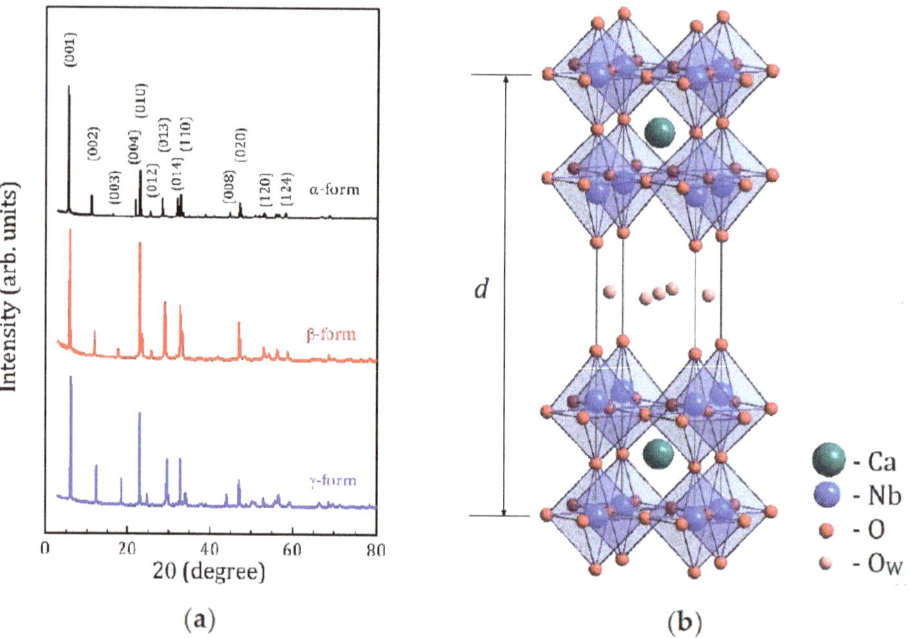

Figure 1. XRD powder patterns for the studied forms of $HCa_2Nb_3O_{10} \cdot yH_2O$ with different hydration levels (**a**) and the unit cell of $HCa_2Nb_3O_{10} \cdot yH_2O$ (**b**) with possible sites for the water oxygen O_w.

Table 1. Lattice parameters (a, c) and unit cell volume (V) for the studied forms of $HCa_2Nb_3O_{10}\cdot yH_2O$.

Parameters	α-Form	β-Form	γ-Form
a (Å)	3.86517(3)	3.86550(9)	3.89267(8)
c (Å)	16.2627(2)	15.1125(7)	14.5254(4)
V (Å^3)	242.957(6)	225.812(18)	220.102(14)

The thermogravimetric (TG) curves that represent the mass decay due to the water release are shown in Figure 2a. As can be seen from the thermogravimetric curves, α-$HCa_2Nb_3O_{10}\cdot yH_2O$ exhibits behavior typical for a low-stable highly hydrated protonated form. Its thermal decomposition proceeds in two main steps, which is typical of protonated layered perovskite-like oxides [40–42]. The first step (T < 373 K) is associated with the release of the intercalated water and the formation of a dehydrated protonated compound:

$$HCa_2Nb_3O_{10}\cdot yH_2O \rightarrow HCa_2Nb_3O_{10} + yH_2O. \tag{1}$$

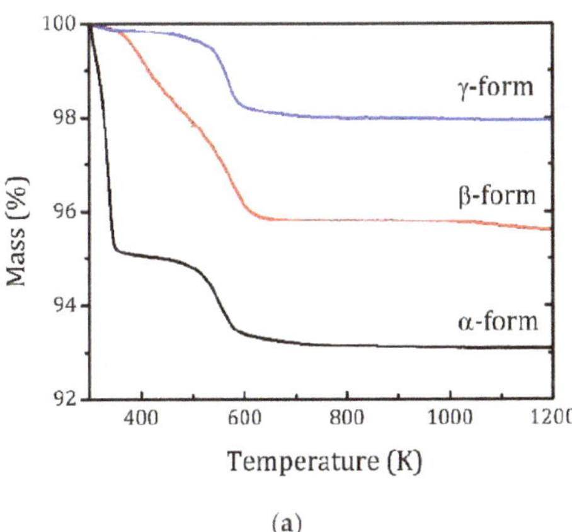

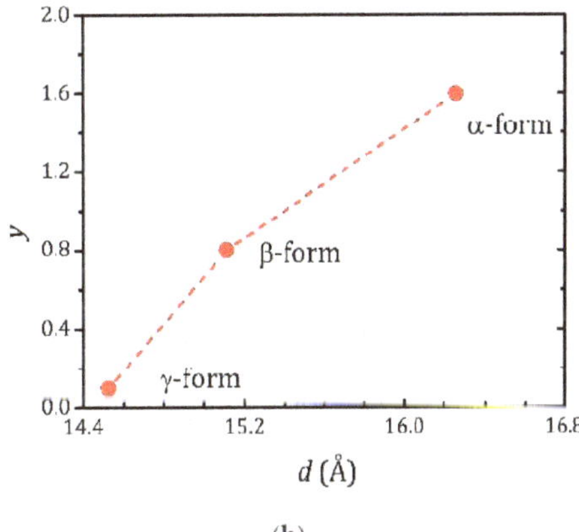

(a)

(b)

Figure 2. (a) TG curves for the studied forms of $HCa_2Nb_3O_{10}\cdot yH_2O$ with different hydration levels; (b) number of H_2O molecules per formula units (y) versus the interlayer space (d).

The second step of the mass loss that occurs at about 525 ÷ 550 K is related with thermal degradation, or so-called topochemical condensation:

$$HCa_2Nb_3O_{10} \rightarrow Ca_2Nb_3O_{9.5} + 0.5H_2O. \tag{2}$$

The thermal decomposition of the γ-form demonstrates similar trends, but the mass loss at the first step is essentially low due to the much lower content of the intercalated water. Thermolysis of the β-form appears to be a more complex process, including gradual evolution of interlayer water in the temperature range of 373 ÷ 525 K, with the subsequent topochemical condensation of the protonated compound. The absence of the mass loss for the β-form at the beginning part of the TG curve indicates its greater thermal stability in comparison with the α-form.

According to the TG analysis, all the studied forms are fully protonated compounds with a substitution degree of K^+ cations for protons H^+ close to 100%. The water content as determined from TG curves results in 1.6, 0.8, and 0.1 H_2O molecules per formula unit for α-, β-, and γ-forms, respectively. When describing layered structures, an important parameter is the interlayer distance d, the distance between the center of the adjacent perovskite slabs. For the studied structures, $d = c$; Figures 1b and 2b show the correlation between the d parameter and the water content, which confirms that water molecules are located within the interlayer space.

2.2. *^{1}H MAS NMR Study*

Figure 3 shows the ^{1}H MAS NMR spectra for the studied forms of $HCa_2Nb_3O_{10} \cdot yH_2O$ acquired at 259 K. As one can see, depending on the hydration level of $HCa_2Nb_3O_{10} \cdot yH_2O$, the spectra differ from each other by the number of spectral lines, their position, and the linewidths. This shows the presence of different proton-containing species in the α-, β-, and γ-forms and their different mobilities.

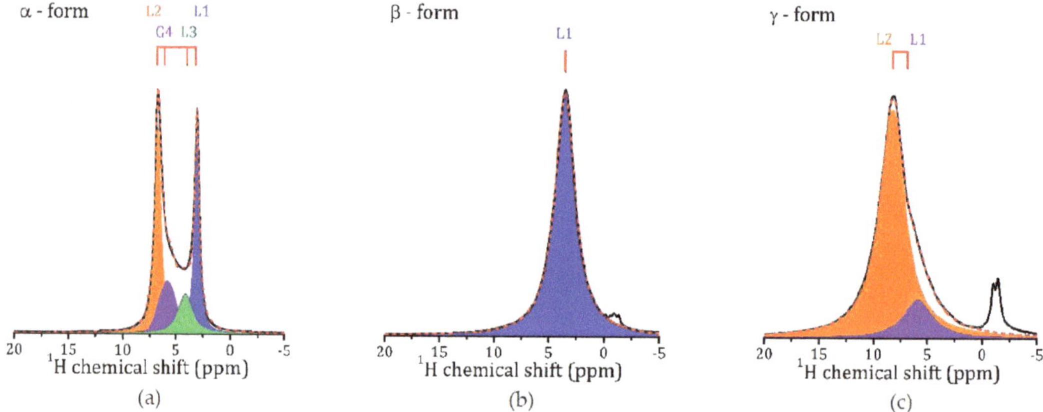

Figure 3. ^{1}H MAS NMR spectra in the α- (**a**), β- (**b**), and γ-forms (**c**) of $HCa_2Nb_3O_{10} \cdot yH_2O$ at 297 K.

At room temperature (297 K), the ^{1}H spectrum of α-$HCa_2Nb_3O_{10}$ (Figure 3a) consists of two narrow intense Lorentzian lines at 3.1 and 6.8 ppm, L1 and L2, respectively, and two lines of lower intensities: Lorentzian line L3 at about 4.1 ppm and Gaussian line G4 at 6.0 ppm. For β-$HCa_2Nb_3O_{10}$ (Figure 3b), it consists of only one rather broad Lorentzian line at 3.6 ppm, whereas for γ-$HCa_2Nb_3O_{10}$ (Figure 3c), the main signal is observed at 8.2 ppm (L2), with a shoulder at 5.9 ppm (L1) (a signal at about -2 ppm can be associated with surface defects and is not discussed further).

To assign the spectral lines to the H-containing species, the evolution of the proton spectra with the temperature decreasing was studied; see Figure 4. As temperature decreases, the spectral lines broaden, and a redistribution of line intensities occurs. Let us first discuss the temperature evolution of the ^{1}H MAS NMR spectrum of γ-form, $HCa_2Nb_3O_{10} \cdot 0.1H_2O$, which is characterized by the lowest water content. At room temperature, the contribution of the L2 line dominates, the relative intensity of L1 is of about 10%, and with sample cooling the line broadens and then disappears. Below 259 K, only the L2 line presents, and with the temperature further decreasing it splits into two lines: Lorentzian type at 8.9 ppm and Gaussian type at 7.8 ppm; see Figure 5. The temperature evolution of the spectral line parameters, namely the isotropic chemical shift (δ_{iso}), the full width at half maximum ($\Delta\nu_{1/2}$), and the relative integral intensities are shown in Figure 6a–c, respectively.

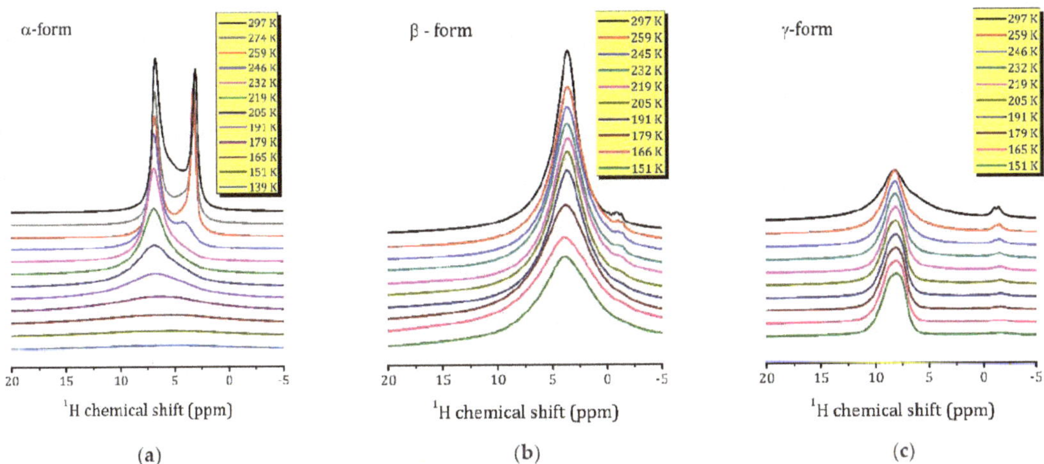

Figure 4. ^{1}H MAS NMR spectra in the α- (**a**), β- (**b**), and γ-forms (**c**) of $HCa_2Nb_3O_{10} \cdot yH_2O$ with the temperature decreasing.

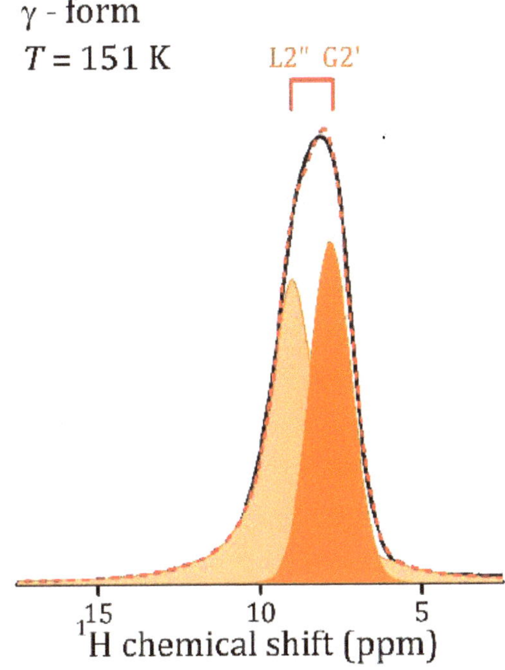

Figure 5. ^{1}H MAS NMR spectra at 151 K in γ-$HCa_2Nb_3O_{10} \cdot yH_2O$ and its decomposition.

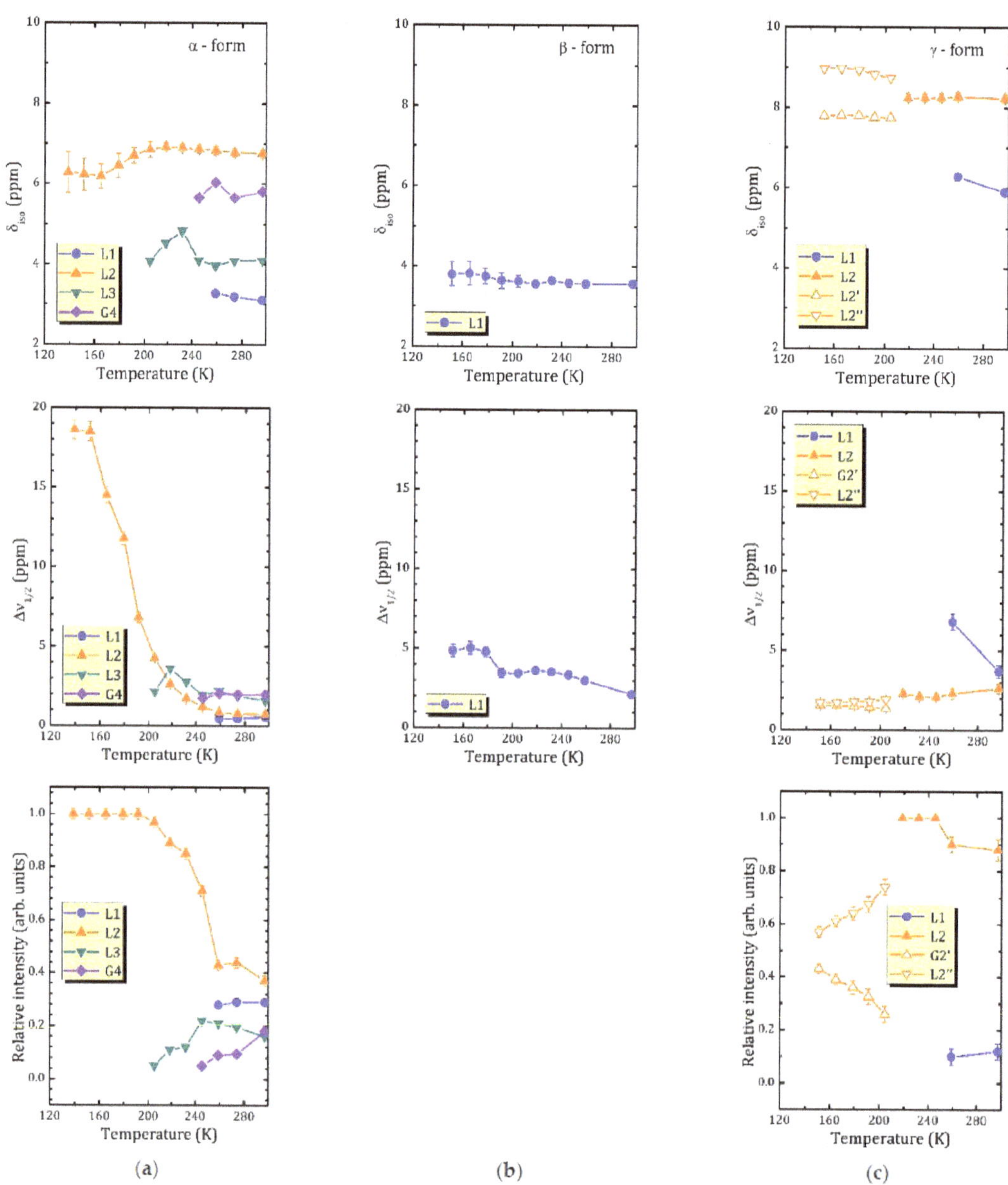

Figure 6. Temperature dependencies of the ^{1}H isotropic chemical shift (the upper row), the line width at half maximum (the middle row), and the integral intensities (the bottom row) for the α- (**a**), β- (**b**), and γ- (**c**) forms of $HCa_2Nb_3O_{10} \cdot yH_2O$.

Based on the TG analysis, one can attribute L1 line to the bulk water. Normally its signal is expected at 5.5 ppm [22], but in a charged nanoconfinement it can be shifted towards a higher frequency. Its contribution is low with the temperature decreasing because of the slowing down of the molecular motion; thus, the line becomes too broad to be resolved. The most intensive line, L2, at about 8 ppm can be associated with the lattice protons in regular sites; e.g., in Ruddlesden–Popper phase $H_2La_2Ti_3O_{10}\cdot0.13H_2O$, the signal of isolated H^+ was reported at 11–13 ppm [13,37]. The splitting of the line at low temperatures may point to two inequivalent cation positions. It is worth noting that down to 151 K, the linewidth of the spectral lines is almost unchanged. This indicates that within the studied temperature range, the proton mobility (translational diffusion) does not change significantly.

The ^1H spectra of α-$HCa_2Nb_3O_{10}$ exhibits the most dramatic changes with temperature: the high field part of the spectrum rapidly disappears with cooling; see Figure 4a. The temperature changes of the spectral parameters are plotted in Figure 6a. As one can see, with the temperature decreasing, the intensities of the spectral lines L1 and G1 rapidly drop, and after cooling down below 245 K, only L2 and L3 remain. Below 200 K, only the L2 peak is visible. Such a complex temperature behavior of the ^1H MAS NMR spectrum of the α-form, as well as its structure, reflect (i) the variety of types of interlayer proton-containing species due to the high content of intercalated water in comparison with the other studied forms, and (ii) the non-obvious mechanisms of interaction between them in a charged environment. Interestingly, in α-form there is no signal associated with isolated protons. Moreover, despite a rather high water content, the only signal that can be associated with the bulk water, the line G4 at 6.1 ppm, has a very low intensity and, similar to the γ-form, rapidly disappears with cooling. Altogether, this suggests the presence of charged water complexed like $H^+ \dots xH_2O$.

According to Ref. [43], the ^1H chemical shift of H_3O^+ ($x = 1$), calculated for water solutions of mineral acids, is expected at 13.3 ppm. With x increasing, the ^1H chemical shift decreases, e.g., for $H^+ \dots 2H_2O$ it was predicted at 8.0 ppm. Our calculations carried out for isolated complexes give 7.3 and 4.6 (17.5) ppm for the isotropic chemical shift for free H_3O^+ and $H_5O_2^+$ clusters, respectively (the number in parenthesis corresponds to the central proton). These calculations are supported by several experimental studies of hydrated layered oxides, in which the signal at 8–11 ppm was assigned to the H_3O^+ [32,36,44–46]. Hence, following both theoretical and experimental studies of other complex layered oxides, and accounting that for α-form of $HCa_2Nb_3O_{10}\cdot yH_2O$ there are 1.6 H_2O molecules and one interlayer proton per one formula, it can be suggested that one water molecule participates in the formation of H_3O^+, the signal L2 at about 7 ppm, whereas other signals correspond to water molecules that are localized in different sites of the charged interlayer space or are part of the more extended charged complexes, like $H^+ \dots 2H_2O$.

The temperature behavior of the L2 linewidth, Figure 6a, is typical for solids [13,47,48] and reflects the slowing down of the molecular motion. Using the onset temperature of motional narrowing, $T_{MN} = 150$ K, one can estimate the activation energy of the line narrowing process within the semi-empirical Waugh-Fedin expression [49]:

$$E_a(\text{eV}) \approx 1.61 \times 10^{-3} \cdot T_{MN}(\text{K}). \tag{3}$$

This results in $E_a \approx 0.24(2)$ eV.

The ^1H MAS NMR spectrum of the β-form consists of one line centered at about 3.6 ppm, which almost does not shift within the studied temperature range; see Figure 5. Taking into account that, according to the TG analysis, the β-form contains 0.8 H_2O molecules per formula unit, and hence per interlayer cation H^+, and no signal from H_3O^+ or H^+ is observed, one can suppose that this line is the result of an exchange between the lattice protons (an expected signal at about 8 ppm as in the γ-form) and the non-hydrogen-bounded water (an expected signal at about 0.8 ppm).

2.3. $^1H\ T_{1\rho}$ Study

To elucidate dynamic processes for all the studied compounds, the temperature dependencies of spin lattice relaxation times in the rotating frame, $T_{1\rho}$ were measured. Relaxation measurements are more sensitive to changes in molecule dynamics than spectroscopic ones [47,50]. For the studied systems, the NMR relaxation is issued mainly by fluctuating strengths of ^{1}H–^{1}H dipole coupling. The latter, being dependent on the relative position of the interacting nuclear spins, is altered by motional processes. As a result, this leads to fluctuations of the Larmor frequency. This process can be described through a correlation function, $G(t)$:

$$G(t) \,=\, \langle \Delta\omega(0) \cdot \Delta\omega(t) \rangle \,=\, G(0) \cdot g(t) \tag{4}$$

where the brackets represent the ensemble average; $g(t)$ contains information about dynamic processes, and its exact expression depends on the spin interaction and diffusion mechanism; $G(0)$ is determined by the mutual nuclear spin arrangement.

Commonly, to describe relaxation processes one uses a spectral relaxation function $j(\omega)$, which is a Fourier-transformed correlation function, $g(t)$. In terms of $j(\omega)$, the dipole contribution to NMR spin-lattice relaxation time T_1, a characteristic time for magnetization recovery after a perturbating pulse, can be written as follows:

$$1/T_1 = G(0) \cdot \left[\frac{1}{3} j(\omega_0) \,+\, \frac{4}{3} j(2\omega_0) \right], \tag{5}$$

where ω_0 is the ^{1}H NMR frequency. By analyzing the temperature dependence of spin lattice relaxation within an appropriate model, one can extract parameters of molecular motion, such as activation energies and correlation times. However, solids normally exhibit slower dynamics as compared to liquids. For such systems, the spin-locking technique is much more fruitful: by applying a locking field ω_1, one can shift the minimum of the temperature dependence of the spin-lattice relaxation time towards the lower temperature in such a way that it falls within the measured temperature range [51–55]. More details can be found in Ref. [13]. At condition $\omega_1 \ll \omega_0$, the relaxation time can be written as

$$1/T_{1\rho} = G(0) \cdot \left[\frac{1}{2} j(2\omega_1) \,+\, \frac{5}{6} j(\omega_0) \,+\, \frac{1}{3} j(2\omega_0) \right]. \tag{6}$$

For $HCa_2Nb_3O_{10} \cdot yH_2O$, the upper limit of temperature is restricted by the water desorption, which according to the TG analysis (Figure 2a) occurs at $T > 300$ K. Application of the spin-locking technique helps to determine the spin motion parameters in a more accurate way.

The relaxation times $T_{1\rho}$ for the studied forms of $HCa_2Nb_3O_{10} \cdot yH_2O$ plotted versus inverse temperature are shown in Figure 7a–c. It should be noted that for all the studied forms within the experimental temperature range, the magnetization recovery is mainly described by a two-exponential function, with characteristic spin-lattice relaxation times $T_{1\rho}{}'$, $T_{1\rho}{}''$ differing from each other in one order of magnitude, except α-form, in which a mono-exponential behavior was observed above 200 K. Examples of the magnetization recovery curves (mono and two-exponential) are shown in Figure 8.

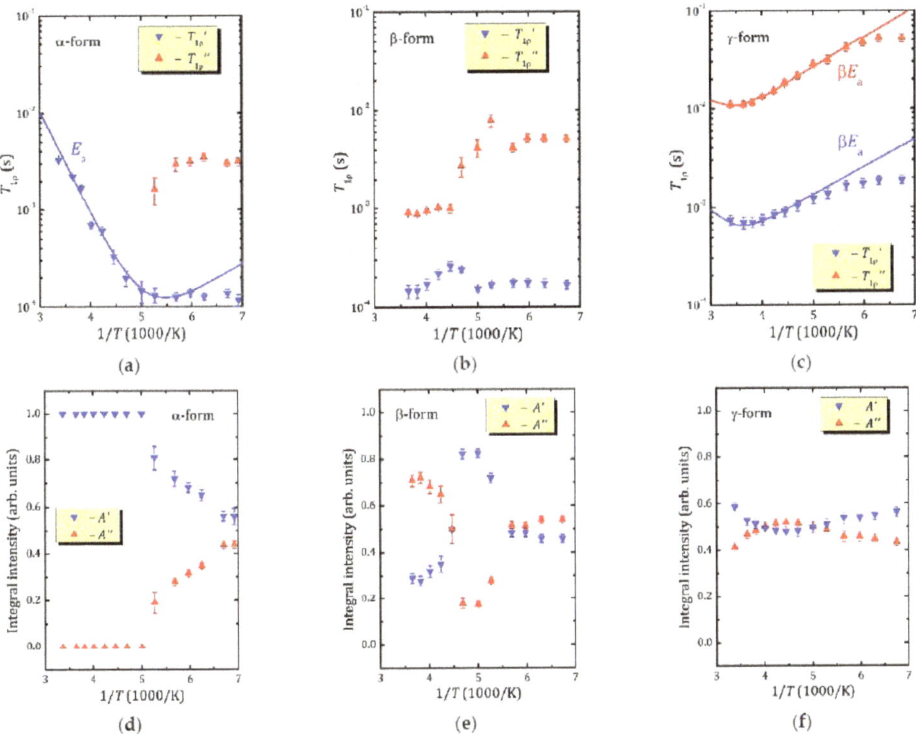

Figure 7. ^{1}H relaxation times $T_{1\rho}{}'$, $T_{1\rho}{}''$ (**a–c**) and their relative contributions A', A'' to the magnetization recovery (**d–f**) for α- (**a,d**), β- (**b,e**), and γ- (**c,f**) forms of $HCa_2Nb_3O_{10} \cdot yH_2O$ versus inverse temperature. The solid lines show the fitting using the KWW correlation function.

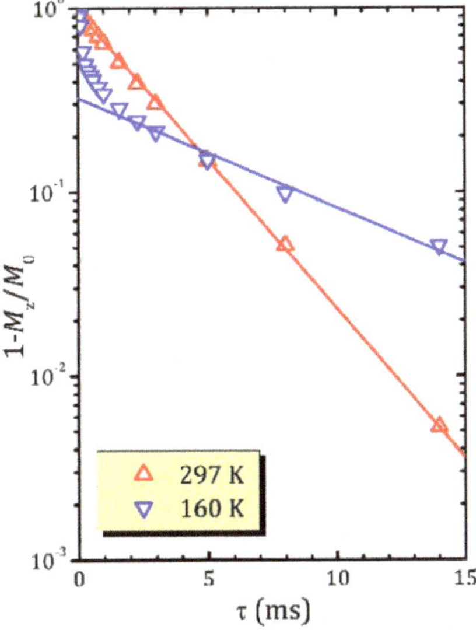

Figure 8. ^{1}H magnetization recovery curves for the α-form of $HCa_2Nb_3O_{10} \cdot yH_2O$ at 297 and 167 K with the exponential fit (solid lines); for 167 K the line corresponds to the slow component only.

As it is clearly seen from Figure 7, depending on the hydration level, $HCa_2Nb_3O_{10}\cdot yH_2O$ demonstrates rather different $T_{1\rho}(1/T)$ behaviors. Let us first discuss the γ-form. The temperature dependence of $T_{1\rho}$ for the least hydrated form of $HCa_2Nb_3O_{10}\cdot yH_2O$ exhibits features similar to $H_{1.83}K_{0.17}La_2Ti_3O_{10}\cdot 0.17H_2O$ [13]. However, it should be noted that the applied locking field was not sufficient to displace the minimum in the middle of the studied temperature range. This complicates the analysis of the experimental data, but the higher pulse power would heat the system excessively.

To determine the proton motion parameters, we used the Kohlrausch-Williams-Watts (KWW) model [56–58] successfully applied to $H_{1.83}K_{0.17}La_2Ti_3O_{10}\cdot 0.17H_2O$ [13]. Commonly, the relaxation in isotropic systems like liquids is described by the well-known Bloembergen-Purcell-Pound (BPP) model [59], which supposes that the exponential function $g(t)$ is as follows:

$$g(t) = e^{-|t|/\tau_c} \tag{7}$$

and that the correlation time τ_c obeys the Arrhenius law: $\tau_c = \tau_0 \exp\left(\frac{E_a}{k_B T}\right)$. Here E_a is the activation energy of hydrogen motion, k_B is the Boltzmann constant, and τ_0 is a pre-exponential factor. The function in Equation (7) results in the following form of the spectral density:

$$j(\omega) = \frac{2\tau_c}{1 + (\omega\tau_c)^2}. \tag{8}$$

This model can be applied to solids, e.g., to describe the translational motion of hydrogen in metallic lattice, but requires some corrections to account for activation energy distribution [47,48], contribution of conduction electrons [60–62], and an exchange between different fractions [48,61,63].

In anisotropic systems, such as ionic conductors, the $T_1(1/T)$ dependence is asymmetric, and a stretched exponential KWW correlation function $g(t) = e^{-(|t|/\tau_s)^\beta}$ is more appropriate for its description. This means that the motion is correlated. These cooperative effects, similar to conduction electrons, contribute mainly at low temperatures, and the corresponding slope is reduced by β [55–58,64]. The spectral density function in this case can be written as

$$j(\omega) = \frac{2\tau_c}{1 + (\omega\tau_c)^{1+\beta}}, \tag{9}$$

with the stretching exponent β ranging from 0 to 1.

As was mentioned above, due to the system limitations for the γ-form of $HCa_2Nb_3O_{10}\cdot yH_2O$, one cannot observe the high temperature slope of the $T_{1\rho}(1/T)$ (Figure 7c), and hence one cannot determine correctly the stretching exponent. That is why to estimate the activation energy we used the parameter $\beta = 0.28$, as determined for $H_{1.83}K_{0.17}La_2Ti_3O_{10}\cdot 0.17H_2O$ [13].

As seen from Figure 7c, the fast ($T_{1\rho}'$) and slow ($T_{1\rho}''$) components of $T_{1\rho}$ exhibit very similar temperature dependencies. Moreover, within the studied temperature range, their contributions are almost tantamount. The fitting within the KWW model results in the very close values of E_a and τ_0: $\{E_a' = 0.223(2)$ eV, $\tau_0' = 8.8(5) \times 10^{-10}$ s$\}$ and $\{E_a'' = 0.213(4)$ eV, $\tau_0'' = 7.8(3) \times 10^{-10}$ s$\}$ for the slow and fast components, respectively. Accounting for the ^1H MAS NMR data, one can suggest that these lines correspond to the isolated interlayer H^+ ions or those in the vicinity of the water molecules.

For the most hydrated α-form above 200 K, the magnetization recovery is described by a single exponent, and the relaxation time $T_{1\rho}$ rapidly decreases with temperature decreasing; see Figure 7a. However, as was mentioned above, below 200 K the character of the magnetization recovery changes, and a second exponent with a longer relaxation time appears. With further temperature decreases, the values of the both short ($T_{1\rho}'$) and slow ($T_{1\rho}''$) components do not change much; nevertheless, the contribution of the $T_{1\rho}''$ component becomes more important and achieves about 44% at 145 K; see Figure 7d. It should be noted that, according to ^1H MAS NMR spectra (Figure 5a), below 200 K there is only one spectral line at about 7 ppm. Such temperature dependencies of the both relaxation times and spectral parameters implicitly show the changes in dynamics of proton-containing species at 200 K. To estimate parameters of the proton motion in α-form, we applied the abovementioned KWW model to the high temperature branch of the $T_{1\rho}{}'(1/T)$. This results in the parameters $E_a = 0.210(2)$ eV, $\tau_0 = 9.0(1) \times 10^{-12}$ s, which can be associated with the translational motion of H_3O^+. Although above 200 K the magnetization recovery curve is described by a single exponential decay, ^{1}H MAS NMR spectra exhibit the existence of different hydrogen-containing species in α-$HCa_2Nb_3O_{10}\cdot yH_2O$ (H_3O^+, H_2O and other); see Figure 6a. This suggests a fast exchange between the species involved in the translational motion. Below 200 K, with the slowing down of the translation, other types of motion (e.g., reorientation) became more important.

The relaxation times for the β-form of $HCa_2Nb_3O_{10}\cdot yH_2O$, Figure 7b, show a complex temperature dependence: within the studied temperature range there are at least two characteristic points (224 and 176 K) at which the proton dynamics change. These changes can also be observed in the temperature dependence of the proton linewidth, but they are less significant.

Therefore, the state of protons and water molecules located in the interlayer space, as well as their dynamics, are determined by the level of hydration. It is noteworthy that the formation of H_3O^+ is confirmed for the most hydrated α-form only, in which it is quite mobile even at low temperatures. For the γ-form, water molecules are not involved in the formation of hydronium ions; protons, behaving as lattice cations, can occupy at least two nonequivalent positions and participate in translational motion. The β-form is the most mysterious. The isotropic chemical shift of only the ^{1}H spectral line indicates an exchange between different species, but it is not possible to unambiguously identify these species from the data obtained. The presence of two characteristic points on the temperature dependence of the proton relaxation time indicates that the mechanism of this exchange is temperature-dependent.

3. Materials and Methods

$KCa_2Nb_3O_{10}$, a precursor for further synthesis of the protonic form, was synthesized by the standard ceramic method at normal conditions (air atmosphere and pressure) using CaO, Nb_2O_5, and K_2CO_3 as parent materials. CaO and Nb_2O_5 were taken in amounts to satisfy the stoichiometry of the reaction:

$$4CaO + 3Nb_2O_5 + K_2CO_3 = 2KCa_2Nb_3O_{10} + CO_2 \uparrow . \tag{10}$$

To compensate for losses during calcination, potassium carbonate was taken with a 30% excess. All the components were mixed and ground in a planetary ball mill in n-heptane. The obtained powder was pelletized and then calcined at 800 °C for 12 h. After that, it was ground in an agate mortar, pelletized again, and calcined at 1100 °C for 24 h.

The first hydrated protonated form of $KCa_2Nb_3O_{10}$, named as α-form of $HCa_2Nb_3O_{10}\cdot yH_2O$, was prepared by acid processing of $KCa_2Nb_3O_{10}$, with an excess of 12 M HNO_3 (50 mL per 2.5 g of the oxide), at room temperature for 24 h. The product was centrifuged, washed with 50 mL of water three times to remove acid residues, and dried over CaO for 24 h. Subsequently, it was stored in a humid air atmosphere.

The second hydrated protonated form of $KCa_2Nb_3O_{10}$, named as β-form of $HCa_2Nb_3O_{10}\cdot yH_2O$, was prepared by hydrothermal treatment of the α- form. For this, 0.5 g of the latter was placed in the laboratory autoclave with 35 mL of water (volume filling approximately 80%) and processed at 150 °C for 7 d. The product obtained was centrifuged and dried over CaO for 24 h.

To obtain the dehydrated protonated form of $KCa_2Nb_3O_{10}$, named as γ-form of $HCa_2Nb_3O_{10}\cdot yH_2O$, the α-form was dried in a desiccator with a vacuum pump (about 1×10^{-4} atm) for 24 h.

Powder XRD analysis was done on a Rigaku Miniflex II diffractometer using monochromatic CuK_α radiation ($\lambda = 0.154056$ nm). Diffractograms were recorded in the 2θ range of

3–120° (step width 0.02°). The lattice parameters were calculated in the tetragonal system on the basis of all the reflections observed using DiffracPlus Topas 4.2 software.

TG analysis was carried out using a Netzsch TG 209 F1 Libra thermobalance. Analysis of samples was carried out in the temperature range 30–900 °C at a heating rate of 10 °C/min in an argon stream at a rate of 90 mL/min.

^{1}H NMR experiments were done using a Bruker Avance IIITM 400 MHz solid-state NMR spectrometer (operating with Topspin version 3.2) using a double-resonance 4 mm low-temperature MAS probe. The temperature was changed within a temperature range of 139 to 297 K and controlled with an accuracy of 0.5 K. ^{1}H NMR MAS spectra were recorded at rotor frequency 12 kHz. To measure T_1 relaxation times, the spin-locking technique was applied, with the rf frequency of the locking pulse equal to 40 kHz. Tetramethylsilane (TMS) was used as an external standard.

The ^{1}H magnetic shielding tensor for H_2O, H_3O^+, and $H_5O_2^+$ was calculated using the Gauge-Independent Atomic Orbital (GIAO) method [65] for the geometries optimized at the B3LYP/6-311G level of theory, as implemented in Gaussian 09 [66]. The theoretical isotropic chemical shift was estimated relative to the magnetic shielding tensor in TMS, calculated at the same level of theory.

4. Conclusions

The results of a comprehensive ^{1}H NMR study of three different forms of the layered perovskite-like niobate $HCa_2Nb_3O_{10} \cdot yH_2O$ can be summarized as follows:

- For the most hydrated α-form, $HCa_2Nb_3O_{10} \cdot 1.6H_2O$, ^{1}H MAS NMR spectra reveal the presence of different proton-containing species: H_3O^+, which comprises all the lattice protons (there is no signal that can be associated with H^+ in regular sites), and water molecules that are localized in different sites of the interlayer slab and supposedly participate in the formation of charged complexes like $H^+ \ldots 2H_2O$. With the temperature decreasing, the signal from the water proton is so broad that it is undetectable; the only detectable signal is from H_3O^+. The activation energy determined from the onset temperature of motional narrowing of 0.24(2) eV is in fair agreement with the relaxation data. Application of the KWW model with the stretching exponent β = 0.28 to the $T_{1\rho}(1/T)$ experimental dependence results in the following molecular motion parameters, which can be associated with the translational diffusion of H_3O^+: $E_a = 0.210(2)$ eV, $\tau_0 = 9.0(1) \times 10^{-12}$ s. Below 200 K, with the slowing down of the translational motion, other types of motion, such as reorientation, became more important.
- ^{1}H MAS NMR spectrum of the β-form, $HCa_2Nb_3O_{10} \cdot 0.8H_2O$, within the studied temperature range consists of one line centered at about 3.6 ppm, which is the result of an exchange between lattice protons and the non-hydrogen-bounded water protons. The temperature dependence of the relaxation time evidences the presence of two characteristic points (224 and 176 K), at which proton dynamics changes.
- The proton NMR spectroscopy study of the γ-form, $HCa_2Nb_3O_{10} \cdot 0.1H_2O$, which is characterized by the lowest water content, reveals the presence of bulk water and interlayer H^+ in two inequivalent positions. The temperature dependencies of the spin-lattice relaxation time in the rotating frame treated withing the KWW model with β = 0.28 results in the following parameters of the proton motion: $E_a = 0.217(8)$ eV, $\tau_0 = 8.2(9) \times 10^{-10}$ s.

We expect that the results obtained will clarify the relationship between the hydration level (and hence the type and, possibly, localization of hydrogen-containing species) and the photocatalytic activity of this class of layered materials. Currently, studies of the photocatalytic activity of the different forms of $HCa_2Nb_3O_{10} \cdot yH_2O$ are under evaluation.

Author Contributions: Conceptualization, M.G.S., O.I.S. and I.A.Z.; methodology, M.G.S. and O.I.S.; investigation, D.Y.N., A.O.A., O.I.S., S.A.K. and A.M.; data curation, M.G.S., O.I.S. and E.A.A.; writing—original draft preparation, M.G.S., O.I.S. and E.A.A.; writing—review and editing, M.G.S., O.I.S. and I.A.Z.; visualization, M.G.S. and E.A.A.; supervision, I.A.Z.; funding acquisition, M.G.S., O.I.S. and I.A.Z. All authors have read and agreed to the published version of the manuscript.

Funding: The study was financially supported by Russian Science Foundation (NMR study of hydrated forms was carried out within the framework of the project No. 19-13-00184, synthesis of a number of hydrated compounds, and their basic characterization was carried out within the framework of the project No. 20-73-00027).

Institutional Review Board Statement: Not applicable.

Informed Consent Statement: Not applicable.

Data Availability Statement: The data presented in this study are available on request from the corresponding author.

Acknowledgments: The samples were studied at the Research Park of Saint Petersburg State University, Centre for X-ray Diffraction Studies, Centre of Thermal Analysis and Calorimetry and Interdisciplinary Resource Centre for Nanotechnology, and Centre for Diagnostics of Functional Materials for Medicine, Pharmacology and Nanoelectronics.

Conflicts of Interest: The authors declare no conflict of interest.

Sample Availability: Samples of the compounds are not available from the authors.

References

1. Liu, Y.; Mao, Z.-Q. Unconventional superconductivity in Sr_2RuO_4. *Phys. C Supercond. Its Appl.* **2015**, *514*, 339–353. [CrossRef]
2. Ogino, H.; Sato, S.; Kishio, K.; Shimoyama, J. Relationship between crystal structures and physical properties in iron arsenides with perovskite-type layers. *Phys. Procedia* **2012**, *36*, 722–726. [CrossRef]
3. Moritomo, Y.; Asamitsu, A.; Kuwahara, H.; Tokura, Y. Giant magnetoresistance of manganese oxides with a layered perovskite structure. *Nature* **1996**, *380*, 141–144. [CrossRef]
4. Domen, K.; Kondo, J.N.; Hara, M.; Takata, T. Photo- and mechano-catalytic overall water splitting reactions to form hydrogen and oxygen on heterogeneous catalysts. *Bull. Chem. Soc. Jpn.* **2000**, *73*, 1307–1331. [CrossRef]
5. Wang, X.; Maeda, K.; Thomas, A.; Takanabe, K.; Xin, G.; Carlsson, J.M.; Domen, K.; Antonietti, M. A metal-free polymeric photocatalyst for hydrogen production from water under visible light. *Nat. Mater.* **2009**, *8*, 76–80. [CrossRef]
6. Toda, K.; Kameo, Y.; Kurita, S.; Sato, M. Crystal structure determination and ionic conductivity of layered perovskite compounds $NaLnTiO_4$ (Ln = rare earth). *J. Alloys Compd.* **1996**, *234*, 19–25. [CrossRef]
7. Petrov, A.A.; Melnikova, N.A.; Petrov, A.V.; Silyukov, O.I.; Murin, I.V.; Zvereva, I.A. Experimental investigation and modelling of the Na^+ mobility in $NaLnTiO_4$ (Ln = La, Nd) ceramics. *Ceram. Int.* **2017**, *43*, 10861–10865. [CrossRef]
8. Thangadurai, V.; Shukla, A.; Gopalakrishnan, J. Proton conduction in layered perovskite oxides. *Solid State Ion.* **1994**, *73*, 9–14. [CrossRef]
9. Kudo, A.; Miseki, Y. Heterogeneous photocatalyst materials for water splitting. *Chem. Soc. Rev.* **2009**, *38*, 253–278. [CrossRef]
10. Rodionov, I.A.; Zvereva, I.A. Photocatalytic activity of layered perovskite-like oxides in practically valuable chemical reactions. *Russ. Chem. Rev.* **2016**, *85*, 248–279. [CrossRef]
11. Minich, I.A.; Silyukov, O.I.; Kulish, L.D.; Zvereva, I.A. Study on thermolysis process of a new hydrated and protonated perovskite-like oxides $H_2K_{0.5}Bi_{2.5}Ti_4O_{13} \cdot yH_2O$. *Ceram. Int.* **2019**, *45*, 2704–2709. [CrossRef]
12. Silyukov, O.I.; Minich, I.A.; Zvereva, I.A. Synthesis of protonated derivatives of layered perovskite-like bismuth titanates. *Glas. Phys. Chem.* **2018**, *44*, 115–119. [CrossRef]
13. Shelyapina, M.G.; Nefedov, D.Y.; Kostromin, A.V.; Silyukov, O.I.; Zvereva, I.A. Proton mobility in Ruddlesden–Popper phase $H_2La_2Ti_3O_{10}$ studied by ^1H NMR. *Ceram. Int.* **2019**, *45*, 5788–5795. [CrossRef]
14. Chen, Y.; Zhao, X.; Ma, H.; Ma, S.; Huang, G.; Makita, Y.; Bai, X.; Yang, X. Structure and dehydration of layered perovskite niobate with bilayer hydrates prepared by exfoliation/self-assembly process. *J. Solid State Chem.* **2008**, *181*, 1684–1694. [CrossRef]
15. Guo, T.; Wang, L.; Evans, D.G.; Yang, W. Synthesis and photocatalytic properties of a pPolyaniline-intercalated layered protonic titanate nanocomposite with a p−n heterojunction structure. *J. Phys. Chem. C* **2010**, *114*, 4765–4772. [CrossRef]
16. Shelyapina, M.G.; Silyukov, O.I.; Lushpinskaia, I.P.; Kurnosenko, S.A.; Mazur, A.S.; Shenderovich, I.G.; Zvereva, I.A. NMR study of intercalates and grafted organic derivatives of $H_2La_2Ti_3O_{10}$. *Molecules* **2020**, *25*, 5229. [CrossRef]
17. Wang, Y.; Wang, C.; Wang, L.; Hao, Q.; Zhu, X.; Chen, X.; Tang, K. Preparation of interlayer surface tailored protonated double-layered perovskite $H_2CaTa_2O_7$ with n-alcohols, and their photocatalytic activity. *RSC Adv.* **2014**, *4*, 4047–4054. [CrossRef]
18. Shelyapina, M.G.; Lushpinskaya, I.P.; Kurnosenko, S.A.; Silyukov, O.I.; Zvereva, I.A. Identification of intercalates and grafted organic derivatives of $H_2La_2Ti_3O_{10}$ by multinuclear NMR. *Russ. J. Gen. Chem.* **2020**, *90*, 760–761. [CrossRef]

19. Voytovich, V.; Kurnosenko, S.; Silyukov, O.; Rodionov, I.; Bugrov, A.; Minich, I.; Malygina, E.; Zvereva, I. Synthesis of n-alkoxy derivatives of layered perovskite-like niobate $HCa_2Nb_3O_{10}$ and study of their photocatalytic activity for hydrogen production from an aqueous solution of methanol. *Catalysts* **2021**, *11*, 897. [CrossRef]

20. Asai, Y.; Ariake, Y.; Saito, H.; Idota, N.; Matsukawa, K.; Nishino, T.; Sugahara, Y. Layered perovskite nanosheets bearing fluoroalkoxy groups: Their preparation and application in epoxy-based hybrids. *RSC Adv.* **2014**, *4*, 26932–26939. [CrossRef]

21. Hojamberdiev, M.; Bekheet, M.F.; Zahedi, E.; Wagata, H.; Kamei, Y.; Yubuta, K.; Gurlo, A.; Matsushita, N.; Domen, K.; Teshima, K. New Dion-Jacobson phase three-layer perovskite $CsBa_2Ta_3O_{10}$ and its conversion to nitrided $Ba_2Ta_3O_{10}$ nanosheets via a nitridation-protonation-intercalation-exfoliation route for water splitt. *Cryst. Growth Des.* **2016**, *16*, 2302–2308. [CrossRef]

22. Jacobson, A.J.; Lewandowski, J.T.; Johnson, J.W. Ion exchange of the layered perovskite $KCa_2Nb_3O_{10}$ by protons. *J. Less Common Met.* **1986**, *116*, 137–146. [CrossRef]

23. Jacobson, A.J.; Johnson, J.W.; Lewandowski, J.T. Intercalation of the layered solid acid $HCa_2Nb_3O_{10}$ by organic amines. *Mater. Res. Bull.* **1987**, *22*, 45–51. [CrossRef]

24. Jacobson, A.J.; Johnson, J.W.; Lewandowski, J.T. Interlayer chemistry between thick transition-metal oxide layers: Synthesis and intercalation reactions of $K[Ca_2Na_{n-3}Nb_nO_{3n+1}]$ (3.ltoreq. n.ltoreq. 7). *Inorg. Chem.* **1985**, *24*, 3727–3729. [CrossRef]

25. Schaak, R.E.; Mallouk, T.E. Prying apart Ruddlesden-Popper phases: Exfoliation into sheets and nanotubes for assembly of perovskite thin films for assembly of perovskite thin films. *Chem. Mater.* **2000**, 3427–3434. [CrossRef]

26. Domen, K.; Yoshimura, J.; Sekine, T.; Kondo, J.; Tanaka, A.; Maruya, K.; Onishi, T. A novel series of photocatalysts with an ion-exchangeable layered structure of niobate. *Catal. Lett.* **1993**, *75*, 2159–2162. [CrossRef]

27. Oshima, T.; Ishitani, O.; Maeda, K. Non-sacrificial water photo-oxidation activity of lamellar calcium niobate induced by exfoliation. *Adv. Mater. Interfaces* **2014**, *1*, 2–5. [CrossRef]

28. Sabio, E.M.; Chamousis, R.L.; Browning, N.D.; Osterloh, F.E. Photocatalytic water splitting with suspended calcium niobium oxides: Why nanoscale is better than bulk—A kinetic analysis. *J. Phys. Chem. C* **2012**, *116*, 3161–3170. [CrossRef]

29. Takata, T.; Furumi, Y.; Shinohara, K.; Tanaka, A.; Hara, M.; Kondo, J.N.; Domen, K. Photocatalytic decomposition of water on spontaneously hydrated layered perovskites. *Chem. Mater.* **1997**, *9*, 1063–1064. [CrossRef]

30. Wang, T.H.; Henderson, C.N.; Draskovic, T.I.; Mallouk, T.E. Synthesis, exfoliation, and electronic/protonic conductivity of the Dion-Jacobson phase layer perovsikite $HLa_2TiTa_2O_{10}$. *Chem. Mater.* **2014**, *26*, 898–906. [CrossRef]

31. Oshima, T.; Yokoi, T.; Eguchi, M.; Maeda, K. Synthesis and photocatalytic activity of $K_2CaNaNb_3O_{10}$, a new Ruddlesden-Popper phase layered perovskite. *Dalt. Trans.* **2017**, *46*, 10594–10601. [CrossRef]

32. Essayem, N.; Tong, Y.Y.; Jobic, H.; Vedrine, J.C. Characterization of protonic sites in $H_3PW_{12}O_{40}$ and $Cs_{1.9}H_{1.1}PW_{12}O_{40}$: A solid-state ^1H, ^2H, ^{31}P MAS-NMR and inelastic neutron scattering study on. *Appl. Catal. A Gen.* **2000**, *194*, 109–122. [CrossRef]

33. Raciulete, M.; Papa, F.; Negrila, C.; Bratan, V.; Munteanu, C.; Pandele-Cusu, J.; Culita, D.C.; Atkinson, I.; Balint, I. Strategy for modifying layered perovskites toward efficient solar light-driven photocatalysts for removal of chlorinated pollutants. *Catalysts* **2020**, *10*, 637. [CrossRef]

34. Arribart, H.; Piffard, Y. Indication from NMR of Grotthuss mechanism for proton conduction in $H_2Sb_4O_{11} \cdot nH_2O$. *Solid State Commun.* **1983**, *45*, 571–575. [CrossRef]

35. Shelyapina, M.G.; Yocupicio-Gaxiola, R.I.; Zhelezniak, I.V.; Chislov, M.V.; Antúnez-García, J.; Murrieta-Rico, F.N.; Galván, D.H.; Petranovskii, V.; Fuentes-Moyado, S. Local structures of two-dimensional zeolites—Mordenite and ZSM-5—Probed by multinuclear NMR. *Molecules* **2020**, *25*, 4678. [CrossRef]

36. Shimizu, T.; Nakai, T.; Deguchi, K.; Yamada, K.; Yue, B.; Ye, J. A visible-light-responsive photocatalyst of nitrogen-doped solid-acid HNb_3O_8-N studied by ultrahigh-field ^1H MAS NMR and $^1H^{93}Nb/^1H^{15}N$ HETCOR NMR in solids. *Chem. Lett.* **2014**, *43*, 80–82. [CrossRef]

37. Cattaneo, A.S.A.S.; Ferrara, C.; Marculescu, A.M.A.M.; Giannici, F.; Martorana, A.; Mustarelli, P.; Tealdi, C. Solid-state NMR characterization of the structure and thermal stability of hybrid organic-inorganic compounds based on a $HLaNb_2O_7$ Dion-Jacobson layered perovskite. *Phys. Chem. Chem. Phys.* **2016**, *18*, 21903–21912. [CrossRef]

38. Krylova, E.A.; Shelyapina, M.G.; Nowak, P.; Harańczyk, H.; Chislov, M.; Zvereva, I.A.; Privalov, A.F.; Becker, M.; Vogel, M.; Petranovskii, V. Mobility of water molecules in sodium- and copper-exchanged mordenites: Thermal analysis and ^1H NMR. *Microporous Mesoporous Mater.* **2018**, *265*, 132–142. [CrossRef]

39. Tani, S.; Komori, Y.; Hayashi, S.; Sugahara, Y. Local environments and dynamics of hydrogen atoms in protonated forms of ion-exchangeable layered perovskites estimated by solid-state ^1H NMR. *J. Solid State Chem.* **2006**, *179*, 3357–3364. [CrossRef]

40. Silyukov, O.; Chislov, M.; Burovikhina, A.; Utkina, T.; Zvereva, I. Thermogravimetry study of ion exchange and hydration in layered oxide materials. *J. Therm. Anal. Calorim.* **2012**, *110*, 187–192. [CrossRef]

41. Utkina, T.; Chislov, M.; Silyukov, O.; Burovikhina, A.; Zvereva, I.A. TG and DSC investigation of water intercalation and protonation processes in perovskite-like layered structure of titanate $K_2Nd_2Ti_3O_{10}$. *J. Therm. Anal. Calorim.* **2016**, *125*, 281–287. [CrossRef]

42. Zvereva, I.A.; Silyukov, O.I.; Chislov, M.V. Ion-exchange reactions in the structure of perovskite-like layered oxides: I. Protonation of $NaNdTiO_4$ complex oxide. *Russ. J. Gen. Chem.* **2011**, *81*, 1434–1441. [CrossRef]

43. Mäemets, V.; Koppel, I. ^{17}O and ^1H NMR chemical shifts of hydroxide and hydronium ion in aqueous solutions of strong electrolytes. *J. Chem. Soc. Faraday Trans.* **1997**, *93*, 1539–1542. [CrossRef]

44. Takagaki, A.; Sugisawa, M.; Lu, D.; Kondo, J.N.; Hara, M.; Domen, K.; Hayashi, S. Exfoliated nanosheets as a new strong solid acid catalyst. *J. Am. Chem. Soc.* **2003**, *125*, 5479–5485. [CrossRef] [PubMed]

45. Jeanneau, E.; Le Floch, M.; Bureau, B.; Audebrand, N.; Louër, D. X-ray diffraction, ^{133}Cs and ^{1}H NMR and thermal studies of $CdZrCs_{1.5}(H_3O)_{0.5}(C_2O_4)_4 \cdot xH_2O$ displaying Cs and water dynamic behavior. *J. Phys. Chem. Solids* **2004**, *65*, 1213–1221. [CrossRef]

46. Grube, E.; Lipton, A.S.; Nielsen, U.G. Identification of hydrogen species in alunite-type minerals by multi-nuclear solid-state NMR spectroscopy. *Phys. Chem. Miner.* **2019**, *46*, 299–309. [CrossRef]

47. Skripov, A.V.; Shelyapina, M.G. Nuclear magnetic resonance. In *Neutron Scattering and Other Nuclear Techniques for Hydrogen in Materials*; Fritzsche, H., Huot, J., Fruchart, D., Eds.; Springer International Publishing: Cham, Switzerland, 2016; pp. 337–376. ISBN 978-3-319-22792-4.

48. Shelyapina, M.G.; Skryabina, N.E.; Surova, L.S.; Dost, A.; Ievlev, A.V.; Privalov, A.F.; Fruchart, D. Proton NMR study of hydrogen mobility in $(TiCr_{1.8})_{1-x}V_x$ hydrides. *J. Alloys Compd.* **2019**, *778*, 962–971. [CrossRef]

49. Waugh, J.S.; Fedin, E.I. Determination of hindered-rotation barriers in solids. *Sov. Phys. Solid State* **1963**, *4*, 1633–1636.

50. Chizhik, V.I.; Chernyshev, Y.S.; Donets, A.V.; Frolov, V.V.; Komolkin, A.V.; Shelyapina, M.G. *Magnetic Resonance and Its Applications*; Springer International Publishing: Cham, Switzerland, 2014; ISBN 9783319052984.

51. Ailion, D.; Slichter, C.P. Study of ultraslow atomic motions by magnetic resonance. *Phys. Rev. Lett.* **1964**, *12*, 168–171. [CrossRef]

52. Slichter, C.P.; Ailion, D. Low-field relaxation and the study of ultraslow atomic motions by magnetic resonance. *Phys. Rev.* **1964**, *135*, A1099–A1110. [CrossRef]

53. Look, D.C.; Lowe, I.J. Nuclear magnetic dipole-dipole relaxation along the static and rotating magnetic fields: Application to gypsum. *J. Chem. Phys.* **1966**, *44*, 2995–3000. [CrossRef]

54. Wolf, D. Effect of correlated self-diffusion on the low-field nuclear-spin relaxation in the rotating reference frame. *Phys. Rev. B* **1974**, *10*, 2724–2732. [CrossRef]

55. Kuhn, A.; Kunze, M.; Sreeraj, P.; Wiemhöfer, H.D.; Thangadurai, V.; Wilkening, M.; Heitjans, P. NMR relaxometry as a versatile tool to study Li ion dynamics in potential battery materials. *Solid State Nucl. Magn. Reson.* **2012**, *42*, 2–8. [CrossRef]

56. Kohlrausch, R. Theorie des elektrischen Rückstandes in der Leidener Flasche. *Ann. Phys.* **1854**, *167*, 179–214. [CrossRef]

57. Williams, G.; Watts, D. Non-symmetrical dielectric relaxation behaviour arising from a simple empirical decay function. *Trans. Faraday Soc.* **1970**, *66*, 80–85. [CrossRef]

58. Williams, G.; Watts, D.C.; Dev, S.B.; North, A. Further considerations of non symmetrical dielectric relaxation behaviour arising from a simple empirical decay function. *Trans. Faraday Soc.* **1971**, *67*, 1323–1335. [CrossRef]

59. Bloembergen, N.; Purcell, E.M.; Pound, R.V. Relaxation effects in nuclear magnetic resonance absorption. *Phys. Rev.* **1948**, *73*, 679–712. [CrossRef]

60. Kasperovich, V.S.; Shelyapina, M.G.; Khar'Kov, B.; Rykov, I.; Osipov, V.; Kurenkova, E.; Ievlev, A.V.; Skryabina, N.E.; Fruchart, D.; Miraglia, S.; et al. NMR study of metal-hydrogen systems for hydrogen storage. *J. Alloys Compd.* **2011**, *509*, S804–S808. [CrossRef]

61. Chizhik, V.I.; Kasperovich, V.S.; Shelyapina, M.G.; Chernyshev, Y.S. Exchange model for proton relaxation in disordered metallic hydrides. *Int. J. Hydrogen Energy* **2011**, *36*, 1601–1605. [CrossRef]

62. Shelyapina, M.G.; Vyvodtceva, A.V.; Klyukin, K.A.; Bavrina, O.O.; Chernyshev, Y.S.; Privalov, A.F.; Fruchart, D. Hydrogen diffusion in metal-hydrogen systems via NMR and DFT. *Int. J. Hydrogen Energy* **2015**, *40*, 17038–17050. [CrossRef]

63. Chizhik, V.I.; Rykov, I.A.; Shelyapina, M.G.; Fruchart, D. Proton relaxation and hydrogen mobility in Ti-V-Cr alloys: Improved exchange model. *Int. J. Hydrogen Energy* **2014**, *39*, 17416–17421. [CrossRef]

64. Cardona, M.; Chamberlin, R.V.; Marx, W. The history of the stretched exponential function. *Ann. Phys.* **2007**, *16*, 842–845. [CrossRef]

65. Ditchfield, R. Self-consistent perturbation theory of diamagnetism I. A gauge-invariant LCAO method for N.M.R. Chemical shifts. *Mol. Phys.* **1974**, *27*, 789–807. [CrossRef]

66. Frisch, M.J.; Trucks, G.W.; Schlegel, H.B.; Scuseria, G.E.; Robb, M.A.; Cheeseman, J.R.; Scalmani, G.; Barone, V.; Mennucci, B.; Petersson, G.A.; et al. *Gaussian 09*; Gaussian, Inc.: Wallingford, CT, USA, 2009.

molecules

MDPI

Article

Screening Metal–Organic Frameworks for Separation of Binary Solvent Mixtures by Compact NMR Relaxometry

Marc Wagemann [1], Natalia Radzik [1], Artur Krzyżak [2] and Alina Adams [1,*]

[1] Institut für Technische und Makromolekulare Chemie, RWTH Aachen University, Templergraben 55, 52056 Aachen, Germany; MaWa0307@web.de (M.W.); natalieradzik@gmail.com (N.R.)

[2] Department of Fossil Fuels, AGH University of Science and Technology, al. Mickiewicza 30, 30-059 Krakow, Poland; akrzyzak@agh.edu.pl

* Correspondence: Alina.Adams@itmc.rwth-aachen.de; Tel.: +49-241-8026428

Abstract: Metal–organic frameworks (MOFs) have great potential as an efficient alternative to current separation and purification procedures of a large variety of solvent mixtures—a critical process in many applications. Due to the huge number of existing MOFs, it is of key importance to identify high-throughput analytical tools, which can be used for their screening and performance ranking. In this context, the present work introduces a simple, fast, and inexpensive approach by compact low-field proton nuclear magnetic resonance (NMR) relaxometry to investigate the efficiency of MOF materials for the separation of a binary solvent mixture. The mass proportions of two solvents within a particular solvent mixture can be quantified before and after separation with the help of a *priori* established correlation curves relating the effective transverse relaxation times $T_{2\mathrm{eff}}$ and the mass proportions of the two solvents. The new method is applied to test the separation efficiency of powdered UiO-66(Zr) for various solvent mixtures, including linear and cyclic alkanes and benzene derivate, under static conditions at room temperature. Its reliability is demonstrated by comparison with results from ^{1}H liquid-state NMR spectroscopy.

Keywords: MOF; separation; binary mixture; low-field NMR relaxometry

check for **updates**

Citation: Wagemann, M.; Radzik, N.; Krzyżak, A.; Adams, A. Screening Metal–Organic Frameworks for Separation of Binary Solvent Mixtures by Compact NMR Relaxometry. *Molecules* **2021**, 26, 3481. https://doi.org/10.3390/molecules26123481

Academic Editor: Robert Brinson

Received: 7 May 2021
Accepted: 5 June 2021
Published: 8 June 2021

Publisher's Note: MDPI stays neutral with regard to jurisdictional claims in published maps and institutional affiliations.

1. Introduction

Metal–organic frameworks (MOFs) are a class of advanced hybrid porous crystalline solids on which metal containing units and organic linkers are held together by strong chemical bonds [1–5]. They have gained increasing interest in recent years as an alternative to traditional nanoporous materials due to their unique combination of high porosities, large surface areas, different pore sizes, and variable topologies tailored through the use of different types of linker and metal nodes [1–5]. Among a large variety of possible applications [6–14], MOFs show great promise for separating and purifying solvent mixtures, processes of key importance in many applications [15–19]. It is expected that they could open an efficient alternative to existing separation procedures, which are expensive, time consuming, and very energy demanding [20,21]. As an example, distillation alone, the workhorse of chemical process industries, is responsible for 10–15% of the world's energy consumption and contributes significantly to the greenhouse gases emissions [20–22]. Therefore, the improvement of current adsorption-based separation methods through the implementation of more efficient adsorbent materials is of growing interest and of fundamental importance to a competitive and environmental friendly technology development.

The separation and purification efficiency of MOFs towards a certain solvent mixture is significantly altered by a variety of factors. They include the nature of the linker and the type of metal which can be chosen to specifically induce a preferential interaction with a certain solvent, the flexibility of the framework, a remarkable property of the MOF materials, as well as the pore size and shape which can be tuned towards size selective sieving [15–19,23–25]. A further factor, which can affect the separation efficiency of certain

The ^{1}H CPMG relaxation decays of the investigated binary solvent mixtures and the corresponding relaxation spectra as obtained by ILT are sensitive to the mass proportion of the two solvents within the mixture, as exemplarily shown by the results depicted in Figure 1. Single relaxation peaks are observed for the mixtures at positions in between those of the pure solvents (Figure 1b). Similar results were obtained for all investigated mixtures (Figure S1, in the Supplementary Materials). The ILT results enable the extraction of an averaged $T_{2\mathrm{eff}}$ for a particular solvent mixture as an efficient parameter for further use.

The variation of the obtained $T_{2\mathrm{eff}}$ of the mixtures with the mass content of the two solvents is exemplarily depicted in Figure 2 for a 1,3,5-triisopropylbenzene (TiPB)/2-pentanone mixture and in Figure S2 (in the Supplementary Materials) for all the other mixtures. All correlation curves could best be described by simple single exponential functions (Figure 2 and Figure S2). Once the correlation curves are established following the measurement and analysis of the CPMG decays, and the fit function for the correlation curve identified, a process which takes around 1 h for each mixture, the separation power of a particular MOF can be quantified within few minutes.

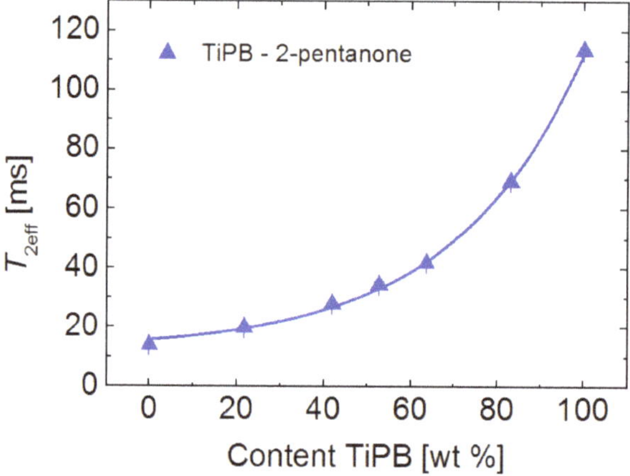

Figure 2. Correlation of the relaxation times with the solvent content in a binary mixture as exemplarily shown for TiPB/2-pentanone. The line depicts the fit result using the equation $T_{2\mathrm{eff\_mixture}} = 3.72 \times \exp(\mathrm{content}/30.3) + 12$ with a correlation factor higher than 0.99. The relaxation times were measured at room temperature.

UiO-66(Zr) (Figure 3) was chosen as a model MOF to test our methodology in the view of its high thermal, mechanical, and chemical stability, well-controlled synthesis procedure, and great promise in the field of separation [23,48–54]. This MOF is built from $Zr_6O_6(CO_2)_{12}$ nodes connected via terephthalate linkers. Its pore system contains tetrahedral and octahedral cavities (free diameters of about 8 Å and 11 Å) [50]. Both cavities are connected by a small triangular window with a pore aperture of about 6.5 Å [50] which needs to be passed by any molecule entering the pores of UiO-66 acting thus as a sieve for larger molecules.

The separation efficiency of powdered UiO-66 due to selective adsorption was tested at room temperature under static conditions for different solvent mixtures by measuring the solvent mixture following equilibration with the MOF and subsequent removal (see the experimental section for more details). The relaxation times of the binary mixtures before and after contact with UiO-66 are different (Figure S3, in the Supplementary Materials). This indicates changes in the mass proportion of the two solvents in the mixture (Figure 4a). In particular, one observes that the content of both n-octane and cyclooctane increases in the mixture with 2-pentanone after separation. This indicates that UiO-66 has a preferential adsorption for 2-pentanone compared to the other two solvents, probably due to a stronger interaction of the 2-pentanone with UiO-66. Furthermore, a partially removal of 2-pentanone from the mixture with TiPB was detected. Given that 2-pentanone should readily enter into UiO-66 and the kinetic diameter of TiPB of about 8.5 Å is much larger than the size of the UiO-66 window, a combination between a sieving mechanism and a blocking of the UiO66 windows by TiPB which prevents that 2-pentanone enters the pores can explain the observed behavior.

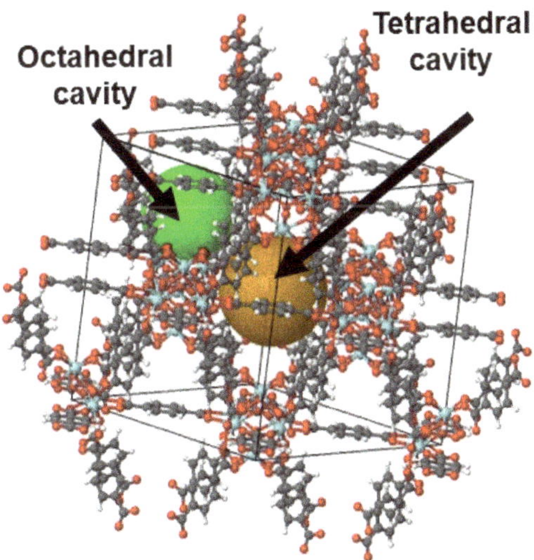

Figure 3. Structure of UiO-66(Zr) showing the carbon (gray), hydrogen (white), oxygen (red), zirconium (blue) and the unit cell (black outline). The tetrahedral and octahedral cavities are indicated by the green and orange spheres. Adapted from [48].

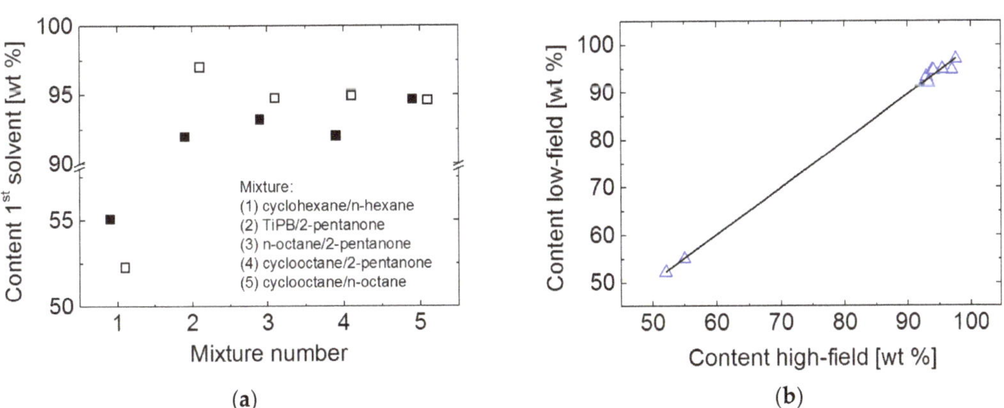

(a)

(b)

Figure 4. (a) Content of the first solvent before (filled symbols) and after (empty symbols) extraction by MOF as calculated based on the T_{2eff} values given in Figure S3. For the n-hexane/cyclohexane a mixture with around 55 wt % cyclohexane was tested. The errors of the obtained values are lower than 0.5%. (b) Correlation of the solvent content in the mixture before and after separation as determined by low-field relaxometry and high-field liquid-state spectroscopy. For the sake of comparison, the results of all mixtures presented in Figure 4a are included. The continuous line is the linear fit result with a correlation factor of 0.997.

The proportion of n-octane and cyclooctane remains largely the same before and after the filtration indicating that the two isomers fail to be separated in the liquid phase under static conditions by powdered UiO-66(Zr). This is an unexpected result because one can assume that n-octane can easily enter into the MOF given that the kinetic diameter of the n-octane is much smaller than the size of the UiO-66 window. Monte Carlo simulations of these two solvents inside the MOF indicate that at zero coverage, n-octane is proportionally distributed between the small and the large cavities and cyclooctane can fit even inside the small, tetrahedral, cavity [52]. However, the simulations give no hint if the cyclooctane, with his kinetic diameter of 8 Å can enter the MOF through the window of 6.5 Å. Given that, to our knowledge, no reports about the separation of the n-octane and cyclooctane by UiO66(Zr) in the liquid phase at room temperature are reported, the raisons behind our observations are not clear. Yet, it has been reported that n-alkanes have the same conformation in a pure liquid state and in a gas state, but they can change conformation in the presence of other solvents [55,56]. Thus, a possible explanation for our observations would be that n-octane changes conformation in the presence of cyclooctane towards a more coil structure and with this its kinetic diameter increases making thus difficult to pass through the MOF window. A further possible explanation of the observed behavior could be related to a solvent induced-breathing process of the MOF. This would lead to an increase in the size of the window such that also cyclooctane could enter the MOF. A similar solvent induced-breathing process had to be taken into account to explain the dependence of the self-diffusion coefficients of methane with the pressure in UiO-66(Zr) [57]. Additional studies are planned for the future to elucidate if a combination of both processes is involved in the observed behavior for this mixture or is largely due only to a change in conformation.

However, a small increase in the content of n-hexane compared to cyclohexane after separation was observed. This is consistent with the reported preferential adsorption of cyclic alkanes over the linear alkanes in UiO-66 measured using vapor phase breakthrough experiments on which the mixtures were diluted using carrier gases such as helium [51,52]. As the pore size of UiO-66 are large enough to accommodate both n-hexane and cyclohexane possible explanations of the much lower separation efficiency are a competitive co-adsorption of both components, the solvent-solvent interaction which in liquid state should play a non-negligible role as well as the particular experimental conditions [58].

The quantified mass proportions of all mixtures by low-field relaxometry approach show an excellent agreement with the results from liquid-state ^{1}H spectroscopy. Proton NMR spectroscopy is another method which can be used to easily quantify the content of certain solvents in a mixture without the need of an a priori calibration or the use of advanced data analysis but requires more sophisticated equipment and for complicated solvent structures measurements at high-magnetic field. Furthermore, the spectra alone fail to give the needed information when applied to certain mixtures including linear alkanes as, for example, for n-hexane/n-octane (Figure 5a). The identical appearance of the spectra of the pure solvents and their mixtures prevents the quantification of the solvent content inside the mixture using solely the differences in the chemical shift of the functional groups. A differentiation between the two solvents and their mixtures can however very easily be achieved with relaxation measurements under the experimental conditions already described (Figure 5b).

The separation of a n-octane/n-hexane mixture by UiO-66 with initially 82 wt % n-octane as determined by weight was also tested. The $T_{2\text{eff}}$ of this mixture of 14.5 ms translates into 81.32 wt % n-octane according to the correlation equation, in good agreement with the weighted value. No separation could be observed for this mixture by UiO-66, probably due to raisons mentioned already before for other binary mixtures.

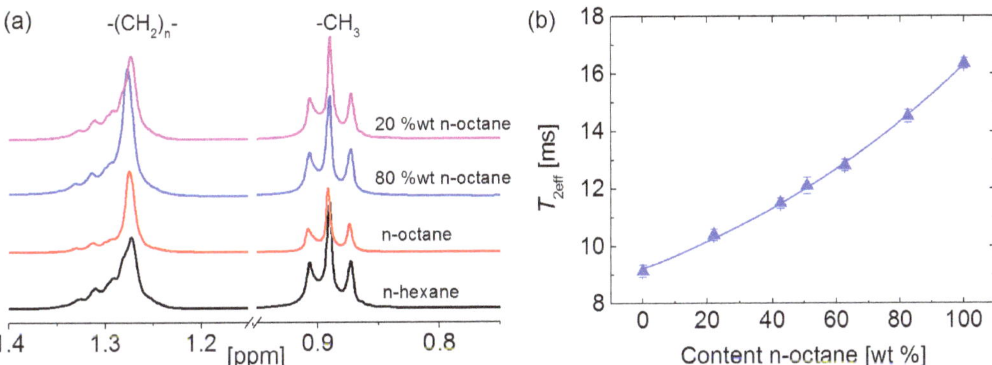

Figure 5. (a) Proton liquid-state spectra of pure n-octane, pure n-hexane, and two mixtures of them measured at room temperature. Both solvents show signals in the same range of chemical shifts. (b) Correlation of the low-field NMR relaxation times with the n-octane content in a binary mixture with n-hexane. The line depicts the fit result using the equation $T_{2eff\_mixture} = 5.03 \times \exp(\text{content}/113) + 4.17$ with a correlation factor higher than 0.99.

3. Materials and Methods

3.1. Samples

All solvents investigated in this study (2-pentanone, n-hexane, cyclohexane, n-octane, cyclooctane, and 1,3,5-triisopropylbenzene (TiPB)) were purchased from Sigma-Aldrich (Germany). Different binary mixtures of the mentioned solvents (cyclooctane/n-octane, cyclooctane/2-pentanone, n-octane/2-pentanone, n-hexane/cyclohexane, and TiPB/2-pentanone) were then prepared by mixing the two solvent components at different mass proportion with a help of a syringe. Due to possible evaporation issues during the preparation of the mixture, the estimated mass proportion was controlled by ^{1}H high-field liquid-state spectroscopy measurements. The binary mixtures were sealed in a glass container in order to keep the concentration constant during the NMR measurement.

The MOF UiO-66(Zr) was purchased from Strem Chemicals (USA). According to the manufacturer, this MOF has a particle size of 0.2–0.5 µm, 1000–1600 m^2 g^{-1} BET, and a pore volume of 0.3–0.5 cm^3 g^{-1} [59]. Following purification at 220 °C overnight under vacuum conditions, the MOF powder was stored under argon atmosphere. The activate MOF retains its crystalline structure as demonstarted by the experimental diffraction patterns (Figure S4, in the Supplementary Materials). All further handling involving MOF samples was also performed under an argon atmosphere to avoid any possible water adsorption from the atmosphere.

To investigate the applicability of proton NMR relaxometry as a fast analytical tool to test the separation power of a certain MOF material, the above mentioned mixtures of the two solvents with a volume proportion of about 1:12 (unless else stated) was used. A total of 65 µL solvent was added for 10 mg of dry MOF. This solvent amount was chosen considering the total amount of solvent that the MOF can uptake. Shortly before starting the measurements, the needed amount of MOF powder was loosely placed in a glass bottle under an argon atmosphere and at room temperature. The mixture of solvents was then gently poured on the top of the MOF using a syringe. The bottle was then immediately closed with a cap sealed with an elastic band and then covered with Teflon coating to prevent the evaporation of hydrocarbons. The prepared system was then left for about 2 h at room temperature to reach an equilibrium state. This equilibration time was confirmed by monitoring the changes in the relaxation times and the ^{1}H spectra of the solvent mixture after different contact times with the metal–organic framework (Figure S5, in the Supplementary Materials). Then a filter was used to remove the MOF and the left solvent mixture was measured by NMR. Each sample was weighed before and after NMR measurements to be sure that no solvent evaporated within the frame of the measurement takes place.

3.2. NMR Experiments

The NMR experiments were performed at room temperature with a single-sided, portable NMR-MOUSE sensor having a static gradient field of about 20 T m^{-1} and working at a proton resonance frequency of 18.2 MHz (Figure 6). A Bruker minispec spectrometer (Germany) was used for pulse generation and signal acquisition. Effective ^{1}H spin–spin relaxation times $T_{2\text{eff}}$ of the pure solvents and their mixtures were determined at room temperature by employing a CPMG pulse sequence [44,45] with an echo time of 0.07 ms. The waiting time between two scans was set for all samples to 4 s in order to avoid heating effects during the measurements. In order to decrease the uncertainty, each measurement was performed three times. The error of the extracted $T_{2\text{eff}}$ values was for all samples less than 1% (see Table S2 for typical values, in the Supplementary Materials). For understanding the trends obtained for the $T_{2\text{eff}}$ values of the pure solvents, literature values, where available, are reported for the corresponding viscosities and self-diffusion coefficients [60–63].

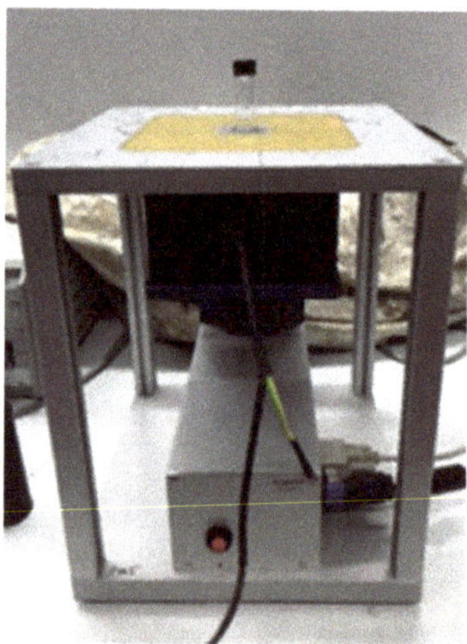

Figure 6. The experimental set-up used to measure the relaxation times of the pure solvents and their mixture before and after the solvent separation by a MOF (here UiO-66(Zr)). The sample to be investigated is simply placed on the top of the profile NMR-MOUSE.

The analysis of all CPMG decays could be best done with the help of a single exponential function for the pure solvents and the mixtures (Figure S1). The distribution of the effective transverse relaxation times was obtained by performing an inverse Laplace transform (ILT) of CPMG relaxation decays.

Proton liquid-state high-field NMR measurements were performed at room temperature using a Bruker Ultrashield magnet operating at a proton frequency of 400 MHz and controlled by an AVANCE 3 console. The pure solvents and the various solvent mixtures were transferred to a 5 mm NMR tube hosting deuterated chloroform. The ^{1}H spectra were acquired after a single 90° radio-frequency pulse with a recycle delay of ten seconds. Evaluation of the mixture ratios was done via integration of component specific peaks. The ^{1}H spectra of all mixtures before and after the separation are depicted in Figures S6–S10 (in the Supplementary Materials) along with the signal assignment and the values of the integral of interest.

4. Conclusions

This work introduces a simple and fast way of quantifying the separation degree of a binary solvent mixture by a MOF material with exemplification on UiO-66(Zr) using a small and low-cost single-sided NMR device. The proposed method is based on correlation curves between the proton effective transverse relaxation times $T_{2\text{eff}}$ and the mass proportions of the two solvents. Once the correlation curves were established, the mass proportions in the filtered mixture can be obtained within a few minutes with great accuracy, as demonstrated by the excellent agreement with the results from liquid-state NMR spectroscopy. The proposed approach can be even applied to characterize solvent mixtures where NMR spectroscopy alone fails. It could help identifying experimental conditions that improve the separation of the mixture components by systematically investigating the impact of particular experimental parameters, such as, e.g., the separation temperature or the presence or absence of a carrier gas. The use of other MOFs with higher separation selectivity for the studied binary mixtures than UiO-66(Zr) would further highlight the potential of this novel method. The whole NMR setup can be introduced inside a synthesis laboratory and it is amenable to automation; thus, helping to save time in searching for adequate separation conditions and MOFs for a particular solvent mixture.

Supplementary Materials: The following are available online. Figure S1: Exemplarily ILT spectra of various investigated mixtures. Figure S2: Correlation curves for different binary mixtures. The continuous lines are the fit results using a single exponential function and has for all samples a correlation factor higher than 0.99. Figure S3: Changes in the effective relaxation times before (closed symbols) and after separation (open symbols) with UiO-66(Zr) for all investigated mixtures. Figure S4: Experimental diffraction patterns of activated UiO-66 (Zr) showing the presence of a fully crystalline structure. Figure S5: Typical ILT of the proton CPMG decays (a) and 1H spectra (b) of the solvent mixture after different mixing times with the metal–organic framework. The solvent-MOF system equilibrates within around 2 h, as no changes between the results can be detected at longer times. Figure S6: 1H high-field spectra of the binary mixture cyclohexane—n-hexane before (top) and after the separation (bottom) by UiO-66(Zr). Figure S7: 1H high-field spectra of the binary mixture TiPB—2-pentanone before (top) and after the separation (bottom) by UiO-66(Zr). The two solvents were mixed in a ratio of about 1:12 as described in the experimental section. Figure S8: 1H high-field spectra of the binary mixture n-octane—2-pentanone before (top) and after the separation (bottom) by UiO-66(Zr). The two solvents were mixed in a ratio of about 1:12 as described in the experimental section. For better view of smaller signals, the signals of n-octane are not shown in full amplitude. Figure S9: 1H high-field spectra of the binary mixture cyclooctane—pentanone before (top) and after the separation (bottom) by UiO-66(Zr). The two solvents were mixed in a ratio of about 1:12 as described in the experimental section. For better view of smaller signals, the signal of cyclooctane is not shown in full amplitude. Figure S10: 1H high-field spectra of the binary mixture cyclooctane—n-octane before (top) and after the separation (bottom) by UiO-66(Zr). The two solvents were mixed in a ratio of about 1:12 as described in the experimental section. For better view of smaller signals, the signal of cyclooctane is not shown in full amplitude. Table S1 (in the Supplementary Materials): Proton transverse relaxation times T2eff of the pure solvents measured using the NMR-MOUSE at room temperature and using the experimental conditions described in the experimental section. Diffusion coefficients and viscosity values are also shown for better understanding the obtained values for the T2eff. Table S2: Reproducibility of the NMR relaxation measurements with exemplification on different mixtures.

Author Contributions: Conceptualization, A.A.; investigation, M.W. and N.R.; resources, A.K. and A.A.; writing—original draft preparation, M.W., N.R. and A.A.; writing—review and editing, A.A.; supervision, A.A.; funding acquisition, A.K. and A.A. All authors have read and agreed to the published version of the manuscript.

Funding: A part of this work was financially supported by National Centre for Research and Development in Poland (contract No. STRATEGMED2/265761/10/NCBR/2015) and by RWTH Aachen University.

Conflicts of Interest: The authors declare no conflict of interest.

References

1. Li, H.; Eddaoudi, M.; O'Keeffe, M.; Yaghi, O.M. Design and synthesis of an exceptionally stable and highly porous metal-organic framework. *Nature* **1999**, *402*, 276–279. [CrossRef]
2. Furukawa, H.; Ko, N.; Go, Y.B.; Aratani, N.; Choi, S.B.; Choi, E.; Yazaydin, A.Ö.; Snurr, R.Q.; O'Keeffe, M.; Kim, J.; et al. Ultrahigh porosity in metal-organic frameworks. *Science* **2010**, *329*, 424–428. [CrossRef]
3. Davis, M.E. Ordered porous materials for emerging applications. *Nature* **2002**, *417*, 813–821. [CrossRef]
4. Ferey, G. Hybrid porous solids: Past, present, future. *Chem. Soc. Rev.* **2008**, *37*, 191–214. [CrossRef]
5. Zhao, D.; Timmons, D.J.; Yuan, D.; Zhou, H.C. Tuning the Topology and Functionality of Metal-Organic Frameworks by Ligand Design. *Acc. Chem. Res.* **2011**, *44*, 123–133. [CrossRef] [PubMed]
6. Eddaoudi, M.; Kim, J.; Rosi, N.; Vodak, D.; Wachter, J.; O'Keeffe, M.; Yaghi, O.M. Systematic design of pore size and functionality in isoreticular MOFs and their application in methane storage. *Science* **2002**, *295*, 469–472. [CrossRef]
7. Czaja, A.U.; Trukhan, N.; Müller, U. Industrial applications of metal–organic frameworks. *Chem. Soc. Rev.* **2009**, *38*, 1284–1293. [CrossRef] [PubMed]
8. Millward, A.R.; Yaghi, O.M. Metal−Organic Frameworks with Exceptionally High Capacity for Storage of Carbon Dioxide at Room Temperature. *J. Am. Chem. Soc.* **2005**, *127*, 17998–17999. [CrossRef] [PubMed]
9. Murray, L.J.; Dincă, M.; Long, J.R. Hydrogen storage in metal–organic frameworks. *Chem. Soc. Rev.* **2009**, *38*, 1294–1314. [CrossRef]
10. Lee, J.; Farha, O.K.; Roberts, J.; Scheidt, K.A.; Nguyen, S.T.; Hupp, J.T. Metal-organic framework materials as catalysts. *Chem. Soc. Rev.* **2009**, *38*, 1450–1459. [CrossRef]
11. Kirchon, A.; Feng, L.; Drake, H.F.; Joseph, E.A.; Zhou, H. From fundamentals to applications: A toolbox for robust and multifunctional MOF materials. *Chem. Soc. Rev.* **2018**, *47*, 8611–8638. [CrossRef]
12. Rasheed, T.; Rizwan, K.; Bilal, M.; Iqbal, H.M.N. Metal-Organic Framework-Based Engineered Materials—Fundamentals and Applications. *Molecules* **2020**, *25*, 1598. [CrossRef]
13. Kuppler, R.J.; Timmons, D.J.; Fang, Q.; Li, J.; Makal, T.A.; Young, M.D.; Yuan, D.; Zhao, D.; Zhuang, W.; Zhou, H. Potential applications of metal-organic frameworks. *Coord. Chem. Rev.* **2009**, *253*, 3042–3066. [CrossRef]
14. Bai, Y.; Dou, Y.; Xie, L.; Rutledge, W.; Li, J.; Zhou, H. Zr-based metal–organic frameworks: Design, synthesis, structure, and applications. *Chem. Soc. Rev.* **2016**, *45*, 2327–2367. [CrossRef]
15. Herm, Z.R.; Wiers, B.M.; Mason, J.A.; van Baten, J.M.; Hudson, M.R.; Zajdel, P.; Brown, C.M.; Masciocchi, N.; Krishna, R.; Long, J.R. Separation of Hexane Isomers in a Metal-Organic Framework with Triangular Channels. *Science* **2015**, *340*, 960–964. [CrossRef] [PubMed]
16. Bloch, E.D.; Queen, W.L.; Krishna, R.; Zadrozny, J.M.; Brown, C.M.; Long, J.R. Hydrocarbon Separations in a Metal-Organic Framework with Open Iron(II) Coordination Sites. *Science* **2012**, *335*, 1606–1610. [CrossRef] [PubMed]
17. Huang, W.; Jiang, J.; Wu, D.; Xu, J.; Xue, B.; Kirillov, A.M. A Highly Stable Nanotubular MOF Rotator for Selective Adsorption of Benzene and Separation of Xylene Isomers. *Inorg. Chem.* **2015**, *54*, 10524–10526. [CrossRef]
18. Herm, Z.R.; Bloch, E.D.; Long, J.R. Hydrocarbon Separations in Metal−Organic Frameworks. *Chem. Mater.* **2014**, *26*, 323–338. [CrossRef]
19. Mukherjee, S.; Desai, A.V.; Ghosh, S.K. Potential of metal–organic frameworks for adsorptive separation of industrially and environmentally relevant liquid mixtures. *Coord. Chem. Rev.* **2018**, *367*, 82–126. [CrossRef]
20. Sholl, D.S.; Lively, R.P. Seven chemical separations to change the world. *Nature* **2016**, *532*, 435–437. [CrossRef]
21. Oak Ridge National Laboratories. *Materials for Separation Technologies: Energy and Emission Reduction Opportunities*; Oak Ridge National Laboratories: Oak Ridge, TN, USA, 2005.
22. Gadalla, M.A.; Olujic, Z.; Jansens, P.J.; Jobson, M.; Smith, R. Reducing CO_2 Emissions and Energy Consumption of Heat-Integrated Distillation Systems. *Environ. Sci. Technol.* **2005**, *39*, 6860–6870. [CrossRef] [PubMed]
23. Wang, X.; Li, L.; Wang, Y.; Li, J.-R.; Li, J. Exploiting the Pore Size and Functionalization Effect in UiO Topology Structures Used for the Separation of Light Hydrocarbons. *Cryst. Eng. Commun.* **2017**, *19*, 1729–1737. [CrossRef]
24. Gutierrez-Sevillano, J.; Calero, S.; Krishna, R. Selective Adsorption of Water from Mixtures with 1-Alcohols by Exploitation of Molecular Packing Effects in CuBTC. *J. Phys. Chem. C* **2015**, *119*, 3658–3666. [CrossRef]
25. Lin, J.M.; He, C.T.; Liu, Y.; Liao, P.Q.; Zhou, D.D.; Zhang, J.P.; Chen, X.M. Metal-organic framework with a pore size/shape suitable for strong binding and close packing of methane. *Angew. Chem. Int. Ed.* **2016**, *55*, 4674–4678. [CrossRef] [PubMed]
26. Henke, S.; Schneemann, A.; Wütscher, A.; Fischer, R.A. Directing the Breathing Behavior of Pillared-Layered Metal–Organic Frameworks via a Systematic Library of Functionalized Linkers Bearing Flexible Substituents. *J. Am. Chem. Soc.* **2012**, *134*, 9464–9474. [CrossRef]
27. Grape, E.S.; Xu, H.; Cheung, O.; Calmels, M.; Zhao, J.; Dejoie, C.; Proserpio, D.M.; Zou, X.; Inge, A.K. Breathing Metal–Organic Framework Based on Flexible Inorganic Building Units. *Cryst. Growth Des.* **2020**, *20*, 320–329. [CrossRef]
28. Shi, Y.X.; Li, W.X.; Zhang, W.H.; Lang, J.P. Guest-Induced Switchable Breathing Behavior in a Flexible Metal–Organic Framework with Pronounced Negative Gas Pressure. *Inorg. Chem.* **2018**, *57*, 8627–8633. [CrossRef]
29. Furukawa, H.; Cordova, K.E.; O'Keeffe, M.; Yaghi, O.M. The chemistry and applications of metal-organic frameworks. *Science* **2013**, *341*, 1230444. [CrossRef]

30. Wilmer, C.E.; Leaf, M.; Lee, C.Y.; Farka, O.K.; Hauser, B.G.; Hupp, J.T.; Snurr, R.Q. Large-scale screening of hypothetical metal-organic frameworks. *Nat. Chem.* **2012**, *4*, 83–89. [CrossRef]
31. Colón, Y.J.; Snurr, R.Q. High-throughput computational screening of metal–organic frameworks. *Chem. Soc. Rev.* **2014**, *43*, 5735–5749. [CrossRef]
32. Chung, Y.G.; Bai, P.; Haranczyk, M.; Leperi, K.T.; Li, P.; Zhang, H.; Wang, T.C.; Duerinck, T.; You, F.; Hupp, J.T.; et al. Computational Screening of Nanoporous Materials for Hexane and Heptane Isomer Separation. *Chem. Mater.* **2017**, *29*, 6315–6328. [CrossRef]
33. Tian, F.; Zhang, X.; Chen, Y. Highly selective adsorption and separation of dichloromethane/trichloromethane on a copper-based metal–organic framework. *RSC Adv.* **2016**, *6*, 31214–31224. [CrossRef]
34. Krishna, R. Screening metal-organic frameworks for mixture separations in fixed-bed adsorbers using a combined selectivity/capacity metric. *RSC Adv.* **2017**, *7*, 35724–35737. [CrossRef]
35. Gutierrez-Sevillano, J.J.; Calero, S.; Krishna, R. Separation of benzene from mixtures with water, methanol, ethanol, and acetone: Highlighting hydrogen bonding and molecular clustering influences in CuBTC. *Phys. Chem. Chem. Phys.* **2015**, *17*, 20114–20124. [CrossRef]
36. Zalesskiy, S.S.; Danieli, E.; Blümich, B.; Ananikov, V.P. Miniaturization of NMR systems: Desktop spectrometers, microcoil spectroscopy, and "NMR on a chip" for chemistry, biochemistry, and industry. *Chem. Rev.* **2014**, *114*, 5641–5694. [CrossRef]
37. Mitchell, J.; Gladden, L.; Chandrasekera, T.; Fordham, E. Low-field permanent magnets for industrial process and quality control. *Prog. Nucl. Mag. Res. Spectrosc.* **2014**, *76*, 1–60. [CrossRef] [PubMed]
38. Adams, A. Analysis of solid technical polymers by compact NMR. *Trends Anal. Chem.* **2016**, *83*, 107–119. [CrossRef]
39. Adams, A.; Kwamen, R.; Woldt, B.; Graß, M. Nondestructive quantification of local plasticizer concentration in PVC by ^{1}H NMR relaxometry. *Macromol. Rapid Commun.* **2015**, *36*, 2171–2175. [CrossRef]
40. Adams, A. Non-destructive analysis of polymers and polymer-based materials by compact NMR. *Magn. Reson. Imaging* **2019**, *56*, 119–125. [CrossRef] [PubMed]
41. Chen, J.J.; Kong, X.; Sumida, K.; Manumpil, M.A.; Long, J.R.; Reimer, J.A. Ex situ NMR relaxometry of metal-organic frameworks for rapid surface-area screening. *Angew. Chem. Int. Ed.* **2013**, *52*, 12043–12046. [CrossRef] [PubMed]
42. Witherspoon, V.J.; Yu, L.M.; Jawahery, S.; Braun, E.; Moosavi, S.M.; Schnell, S.K.; Smit, B.; Reimer, J.A. Translational and Rotational Motion of C8 Aromatics Adsorbed in Isotropic Porous Media (MOF-5): NMR Studies and MD Simulations. *J. Phys. Chem. C* **2017**, *121*, 15456–15462. [CrossRef]
43. Horch, C.; Schlayer, S.; Stallmach, F. High-pressure low-field ^{1}H NMR relaxometry in nanoporous materials. *J. Magn. Reson.* **2014**, *240*, 24–33. [CrossRef] [PubMed]
44. Carr, H.; Purcell, E. Effects of Diffusion on Free Precession in Nuclear Magnetic Resonance Experiments. *Phys. Rev.* **1954**, *94*, 630–638. [CrossRef]
45. Meiboom, S.; Gill, D. Modified Spin-Echo Method for Measuring Nuclear Relaxation Times. *Rev. Sci. Instrum.* **1958**, *29*, 688–691. [CrossRef]
46. Tofts, P.S.; Lloyd, D.; Clark, C.A.; Barker, G.J.; Parker, G.J.M.; McConville, P.; Baldock, C.; Pope, J.M. Test Liquids for Quantitative MRI Measurements of Self-Diffusion Coefficient In Vivo. *Magn. Reson. Med.* **2000**, *43*, 368–374. [CrossRef]
47. Korb, J.-P. Multiscale nuclear magnetic relaxation dispersion of complex liquids in bulk and confinement. *Prog. Nucl. Magn. Reson. Spectrosc.* **2018**, *104*, 12–55. [CrossRef]
48. UiO-66 Metal Organic Framework. Available online: https://www.chemtube3d.com/mof-uio66/ (accessed on 23 April 2021).
49. Cavka, J.H.; Jakobsen, S.; Olsbye, U.; Guillou, N.; Lamberti, C.; Bordiga, S.; Lillerud, K.P. A New Zirconium Inorganic Building Brick Forming Metal Organic Frameworks with Exceptional Stability. *J. Am. Chem. Soc.* **2008**, *130*, 13850–13851. [CrossRef] [PubMed]
50. Biswas, S.; van der Voort, P.A. General Strategy for the Synthesis of Functionalised UiO-66 Frameworks: Characterisation, Stability and CO_2 Adsorption Properties. *Eur. J. Inorg. Chem.* **2013**, 2154–2160. [CrossRef]
51. Barcia, P.S.; Guimaraes, D.; Mendes, P.A.P.; Silva, J.A.C.; Guillerm, V.; Chevreau, H.; Serre, C.; Rodrigues, A.E. Reverse Shape Selectivity in the Adsorption of Hexane and Xylene Isomers in MOF UiO-66. *Microporous Mesoporous Mater.* **2011**, *139*, 67–73. [CrossRef]
52. Duerinck, T.; Bueno-Perez, R.; Vermoortele, F.; De Vos, D.E.; Calero, S.; Baron, G.V.; Denayer, J.F.M. Understanding Hydrocarbon Adsorption in the UiO-66 Metal-Organic Framework: Separation of (Un)saturated Linear, Branched, Cyclic Adsorbates, Including Stereoisomers. *J. Phys. Chem. C* **2013**, *117*, 12567–12578. [CrossRef]
53. Bozbiyik, B.; Duerinck, T.; Lannoeye, J.; De Vos, D.E.; Baron, G.V.; Denayer, J.F.M. Adsorption and separation of n-hexane and cyclohexane on the UiO-66 metal–organic framework. *Microporous Mesoporous Mater.* **2014**, *183*, 143–149. [CrossRef]
54. Usman, M.; Helal, A.; Abdelnaby, M.M.; Alloush, A.M.; Zeama, M.; Yamini, Z.H. Trends and Prospects in UiO-66 Metal-Organic Framework for CO_2 Capture, Separation, and Conversion. *Chem. Rec.* **2021**, *21*, 1–22. [CrossRef]
55. Thomas, L.L.; Christakis, T.J.; Jorgensen, W.L. Conformation of Alkanes in the Gas Phase and Pure Liquids. *J. Phys. Chem. B* **2006**, *110*, 21198–21204. [CrossRef]
56. Nikki, K.; Inakura, H.; Wu-Le; Suzuki, N.; Endo, T. Remarkable changes in conformations of *n*-alkanes with their carbon numbers and aromatic solvents. *J. Chem. Soc. Perkin Trans.* **2000**, 2370–2373. [CrossRef]

57. Yang, Q.; Jobic, H.; Salles, F.; Kolokolov, D.; Guillerm, V.; Serre, C.; Maurin, G. Probing the Dynamics of CO_2 and CH_4 within the Porous Zirconium Terephthalate UiO-66(Zr): A Synergic Combination of Neutron Scattering Measurements and Molecular Simulations. *Chem. Eur. J.* **2011**, *17*, 8882–8889. [CrossRef]
58. Zhao, W.W.; Zhang, C.Y.; Yan, Z.G.; Bai, L.P.; Wang, X.; Huang, H.; Zhou, Y.Y.; Xie, Y.; Li, F.S.; Li, J.R. Separations of substituted benzenes and polycyclic aromatic hydrocarbons using normal- and reverse-phase high performance liquid chromatography with UiO-66 as the stationary phase. *J. Chromatogr. A* **2014**, *1370*, 121–128. [CrossRef] [PubMed]
59. Strem Chemicals. Available online: https://www.strem.com/catalog/v/40-1105/85/zirconium_1072413-89-8 (accessed on 23 April 2021).
60. Fischer, J.; Weiss, A. Transport properties of liquids. V. self-diffusion, viscosity, and mass density of ellipsoidal shaped molecules in the pure liquid phase. *Ber. Bunsenges. Phys. Chem.* **1986**, *90*, 896–905. [CrossRef]
61. Funke, H.H.; Argo, A.M.; Falconer, J.L.; Noble, R.D. Separations of Cyclic, Branched, and Linear Hydrocarbon Mixtures through Silicalite Membranes. *Ind. Eng. Chem. Res.* **1997**, *36*, 137–143. [CrossRef]
62. Chua, L.M.; Hitchcock, I.; Fletcher, R.S.; Holt, E.M.; Lowe, J.; Rigby, S.P. Understanding the Spatial Distribution of Coke Deposition within Bimodal Micro-/Mesoporous Catalysts using a Novel Sorption Method in Combination with Pulsed-gradient Spin echo NMR. *J. Catal.* **2012**, *286*, 260–265. [CrossRef]
63. Van der Perre, S.; Van Assche, T.; Bozbiyik, B.; Lannoeye, J.; De Vos, D.E.; Baron, G.V.; Denayer, J.F.M. Adsorptive Characterization of the ZIF-68 Metal-Organic Framework: A Complex Structure with Amphiphilic Properties. *Langmuir* **2014**, *30*, 8416–8424. [CrossRef]

molecules

Article

Gadolinium-Based Paramagnetic Relaxation Enhancement Agent Enhances Sensitivity for NUS Multidimensional NMR-Based Metabolomics

Chandrashekhar Honrao [1], Nathalie Teissier [1], Bo Zhang [1], Robert Powers [2,3,*] and Elizabeth M. O'Day [1,*]

[1] Olaris, Inc., Waltham, MA 02451, USA; chonrao@olarisbor.com (C.H.); nteissier@olarisbor.com (N.T.); bozhangchem@gmail.com (B.Z.)

[2] Department of Chemistry, University of Nebraska-Lincoln, Lincoln, NE 68588, USA

[3] Nebraska Center for Integrated Biomolecular Communication, University of Nebraska-Lincoln, Lincoln, NE 68588, USA

* Correspondence: rpowers3@unl.edu (R.P.); eoday@olarisbor.com (E.M.O.)

Abstract: Gadolinium is a paramagnetic relaxation enhancement (PRE) agent that accelerates the relaxation of metabolite nuclei. In this study, we noted the ability of gadolinium to improve the sensitivity of two-dimensional, non-uniform sampled NMR spectral data collected from metabolomics samples. In time-equivalent experiments, the addition of gadolinium increased the mean signal intensity measurement and the signal-to-noise ratio for metabolite resonances in both standard and plasma samples. Gadolinium led to highly linear intensity measurements that correlated with metabolite concentrations. In the presence of gadolinium, we were able to detect a broad array of metabolites with a lower limit of detection and quantification in the low micromolar range. We also observed an increase in the repeatability of intensity measurements upon the addition of gadolinium. The results of this study suggest that the addition of a gadolinium-based PRE agent to metabolite samples can improve NMR-based metabolomics.

Keywords: NMR; metabolomics; paramagnetic; relaxation; gadolinium

Citation: Honrao, C.; Teissier, N.; Zhang, B.; Powers, R.; O'Day, E.M. Gadolinium-Based Paramagnetic Relaxation Enhancement Agent Enhances Sensitivity for NUS Multidimensional NMR-Based Metabolomics. *Molecules* **2021**, *26*, 5115. https://doi.org/10.3390/molecules26175115

Academic Editor: Robert Brinson

Received: 22 July 2021
Accepted: 20 August 2021
Published: 24 August 2021

Publisher's Note: MDPI stays neutral with regard to jurisdictional claims in published maps and institutional affiliations.

1. Introduction

Metabolomics is a rapidly expanding field that relies on the detection and quantification of small molecular-weight (MW < 1500 Daltons) compounds present in a biological sample. Metabolite levels are often correlated with different disease states or phenotypic outcomes, which can lead to the development of highly valuable biomarkers and provide novel insights into human health and disease [1–7]. Nuclear magnetic resonance (NMR) spectroscopy has proven to be a powerful tool for metabolomics that meets the analytical requirements needed to achieve a robust and accurate characterization of the metabolome [8–11]. Conventional NMR-based approaches rely on one-dimensional (1D) ^{1}H NMR experiments, which can facilitate the absolute quantification of metabolites. However, chemical shift overlap may limit the number of metabolites that can be accurately measured, which often relies on the application of peak-fitting algorithms. The size and completeness of the reference database used by these peak fitting algorithms will also limit the number of metabolites that can be quantified. Multi-dimensional techniques such as two-dimensional (2D) ^{1}H-^{13}C Heteronuclear Single Quantum Correlation (HSQC) spectroscopy can increase resolution by dispersing the chemical shifts along the carbon dimension, but necessitates long acquisition times due to the low natural abundance of ^{13}C (1.1%), thus limiting the real-world practicality of this approach [12].

Expanding upon the work of Rai [13] and Von Schlippenbach [14], we recently demonstrated that non-uniform sampling (NUS) can be used to reduce the acquisition time of a 2D ^{1}H-^{13}C HSQC experiment to empower semi-quantitative metabolomics [15]. Indeed, a one-hour experiment using a 25% NUS ^{1}H-^{13}C HSQC led to 4-fold improvement in

67

sensitivity, which also yielded highly linear and repeatable data. Further, we established guidelines based on a signal-to-noise ratio (S/N) to enable the reliable detection of a broad range of metabolites in the low micromolar range with a coefficient of variation (CV) of less than 20%. Using these results as our baseline, we sought herein to systematically evaluate the effects of relaxation delays in combination with paramagnetic relaxation enhancement (PRE) agents to further improve the sensitivity of 2D NMR experiments for metabolomics. First, we improved the mean signal intensity and S/N of a 25% NUS ^{1}H-^{13}C HSQC experiment by optimizing the relaxation delay and the number of scans. Then, we observed that the addition of a gadolinium-based PRE agent further improved the S/N of the 25% NUS ^{1}H-^{13}C HSQC spectra for both a model mixture and plasma samples. A lower limit of detection and quantification was achieved for most metabolites, but the most dramatic improvement in signal intensity was seen for the weakest peaks. We also observed that the addition of the PRE agent maintained linearity for all metabolites over a concentration range from 50 μM to 2 mM. These intensity measurements were highly repeatable, leading to smaller CVs. Overall, our results demonstrate that PRE agents can improve the sensitivity of 2D NUS NMR spectra routinely used in metabolomic studies.

2. Results

2.1. Optimizing the Relaxation Delay for Semi-Quantitative Metabolomics

A fundamental principle of NMR spectroscopy is that increasing the number of experimental scans (N) increases the S/N ratio by a factor of $\sqrt{N}$ [16]. For pulsed NMR experiments, the relaxation delay, commonly known as d1, is the time required between scans to allow spins to return to equilibrium. The optimal d1 time depends on the longitudinal relaxation (T_1) rate—the time required for full restoration of the nuclear spin to equilibrium along the direction of the polarizing magnetic field [16]. Each nuclei in a molecule has a different T_1 value, and for small molecules like metabolites T_1 values can be several seconds long. For example, formate has a $T_1 > 9$ s at 600 MHz [17]. For quantitative NMR, it is advised to set d1 to 5 × T_1 of the slowest relaxing nuclei in a sample [17]. This would require a d1 of upwards of a minute in length, leading to impractically long acquisition times that are not feasible for high-throughput NMR metabolomics. In practice, d1 is commonly set to a pre-determined value that allows for a relative quantitative comparison between spectra collected under identical conditions. It is important to note that only a comparison between the same metabolite can be made in this manner across the spectral dataset. A comparison between two or more different metabolites would be meaningless because of the d1-dependent variation in peak integrals that distorts the relationship between peak integral and metabolite concentration.

A model mixture ("Reference 1") was composed of 29 commonly observed human metabolites, which included amino acids, organic acids, biogenic amines, sugars, etc., from the literature [18–21] as well as metabolites commonly observed in our own clinical studies. To find the optimal d1 for a model mixture of 29 metabolites (Reference 1), we recorded time equivalent experiments (4 min ± 8 s) with varying d1 values of 1.5 s, 1.2 s, 0.8 s and 0.6 s and observed the changes in both the 1D ^{1}H NMR spectra and 2D 25% NUS ^{1}H-^{13}C HSQC spectra (Figure 1). At first, as the d1 decreased, the signal intensity for the majority of the metabolites increased, which is expected due to the increased number of scans (N = 64 to 92 for the 1D- and N = 36 to 84 for the 2D-experiments). For the 25% NUS ^{1}H-^{13}C HSQC spectra, we observed a steady increase in the mean intensity of metabolites from 2.9×10^7 to 4.9×10^7 as d1 decreased from 1.5 s to 0.8 s. Similarly, the mean S/N increased from 98.48 to 115.34. However, the mean S/N and intensity reached a maximum at a d1 of 0.8 s. As evident by the expanded regions of the 1D ^{1}H NMR spectra (Figure 1a), peak intensities began to decrease at a d1 of 0.6 s despite the larger number of scans. This is consistent with the 25% NUS ^{1}H-^{13}C HSQC spectra at a d1 of 0.6 s, where the mean S/N and intensity decreased to 106.18 and 4.7×10^7, respectively. Furthermore, significant solvent artifacts were observed in the HSQC spectra relative to longer d1 values. Presumably, at a d1 of 0.6 s, factors related to T_1 dominate spectral sensitivity, which could not be negated by the allowed increase in the number of scans. This led us to select 0.8 s as the optimal d1 value for improved S/N.

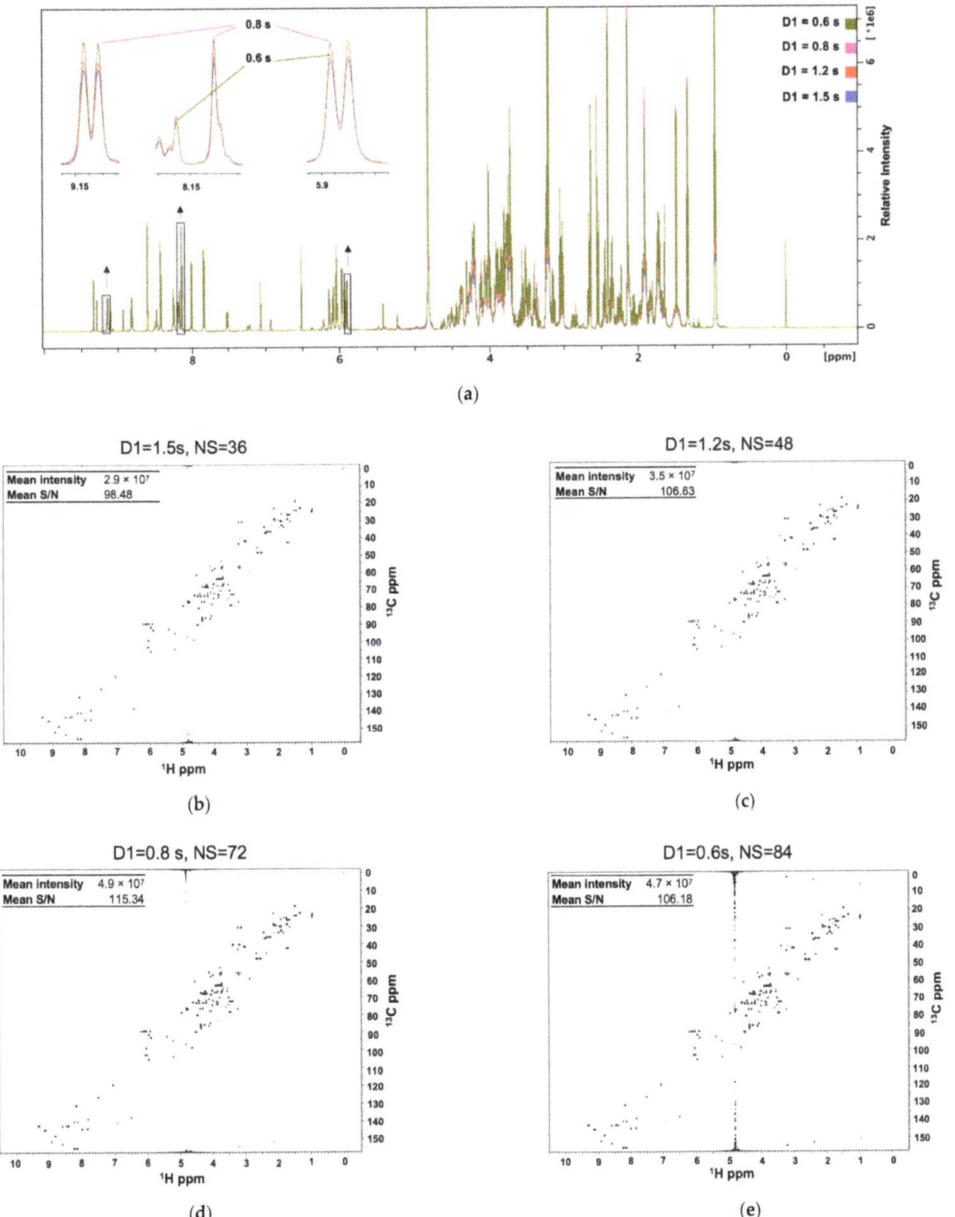

Figure 1. An optimized combination of relaxation delay (d1) and number of scans (N) improved the S/N for (a) 1D ^{1}H and (b–e) 2D ^{1}H-^{13}C HSQC spectra of a model mixture of metabolites in time equivalent experiments. Four 1D ^{1}H NMR spectra are overlaid and color-coded according to the d1 value: 1.5 s (blue), 1.2 s (red), 0.8 s (green), 0.6 s (purple). The boxed regions in the 1D ^{1}H NMR spectra are shown as expanded inserts above each corresponding arrow.

2.2. Gadolinium Provides Enhanced Sensitivity

The addition of PRE agents has been previously used to accelerate NMR data acquisition [13,17,22–25]. PRE agents contain unpaired electrons and decrease T_1 relaxation times for all nuclei in a sample due to dipolar interactions between nuclear and electron spin states. The PRE effect is very large, owing to the large magnetic moment of an unpaired electron, and can be tunable by adjusting the concentration of the PRE agent [17]. By combining NUS with the relaxation enhancing agent, Cu(EDTA), Rai and colleagues demonstrated a 22-fold reduction in the 2D ^{1}H-^{13}C HSQC data collection time to quantify a handful of urine metabolites [13]. Gadolinium-based contrast agents have been widely used in MRI diagnostic imaging, for studying soluble proteins, for characterizing protein-protein, protein-oligosaccharides, and protein-nucleic acid complexes, and for investigating membrane proteins using NMR spectroscopy [26,27]. Sakol et al. have also shown the utility of the Gd-based contrast agent, Gd-DOTA, for cellular localization studies using NMR spectroscopy [28]. Similarly, Mulder and colleagues utilized gadolinium-based PRE agents and achieved a 3- to 4-fold improvement in acquisition time for quantifying several plasma metabolites [17]. We sought to expand upon these findings by focusing on parameters to increase spectral sensitivity for a fixed-time experiment (1 h $\pm$ 4 min) instead of accelerating acquisition times.

We first assessed the 1D ^{1}H spectral changes for Reference 1 (Supplementary Material Table S1) with a d1 of 0.8 s with increasing concentrations (0.25 mM to 1 mM) of Cu (EDTA) and Gadobutrol (Gd) (Figure S1), a gadolinium-containing macrocyclic that has previously been shown to enhance the relaxation rates of urine metabolites [17,25] (Figure S2). In general, contrast agents containing Gd shorten T_1 and T_2 relaxation rates through a dipole–dipole interaction between the unpaired electron of Gd and nuclei in the compound. The decrease in T_1 and T_2 rates depends on the contrast agent used and its concentration, the charge state of the compound, the viscosity of the solution, and the protein affinity of either the compound or contrast agent, among other issues. The typical range of T_1 values for nuclei of common metabolites such as glucose, lactate, citrate, acetate, glutamine, and alanine are between 0.9 and 4 s. Similarly, T_2 values range from 100 to 600 ms [17,28,29]. In the presence of Gd, T_1 values can decrease from 2- to 10-fold depending on the concentration of Gd. A similar reduction is observed for T_2, but is more pronounced at higher Gd concentrations. Accordingly, NMR resonances will significantly broaden into the baseline with the increase in Gd concentration [28]. Experimentally, we observed that a concentration of Gd at 0.25 mM allowed us to decrease our recycle delay to 0.8 s and achieve an overall increase in sensitivity while avoiding substantial line broadening. As the concentration of the Gd agent increased, the decrease in T_2 and the associated peak broadening eventually eclipsed the reduction in T_1 and negated any intensity gains from a larger number of scans [17,28,30,31]. In agreement with these observations, at 0.25 mM Gd, we noted an increase in intensities for the majority of metabolite resonances. As the concentration of Gd increased to 0.5 mM, a handful of metabolite resonances continued to show an increase in intensity, while others began to broaden. At 1 mM Gd, the majority of resonances were diminished compared to the control that lacked Gd. Interestingly, our results are in line with the theoretical optimal recycle delay predictions of Rovnyak et al. [32]. To perform the comparison, we identified NMR relaxation times reported in the literature for metabolites included in our study. For example, the work by Mulder et al. [17] demonstrated that the addition of Gd at a concentration of 0.5 mM to a mixture of small molecules (glucose, creatinine, citrate, glutamine, acetate, alanine, etc.) greatly reduced the T_1 relaxation times by 2- to 10-fold, resulting in an average T_1 relaxation time of ~0.6 s. Using the equation derived from Rovnyak et al., in the presence of Gd the theoretical optimal recycle delay would be ~0.8 s (1.26 $\times$ 0.6 s), which is in perfect agreement with our experimental findings of an optimal d1 of 0.8 s. In the presence of Cu (EDTA), we observed a decrease in NMR resonance intensities and significant line broadening at all concentrations tested. These results suggest that the addition of 0.25 mM of Gd may offer an optimal improvement in S/N. Indeed, when we recorded a 25% NUS ^{1}H-^{13}C HSQC spectrum with a d1 of 0.8 s in the presence of 0.25 mM Gd, we observed an increase in both the mean intensity and mean S/N for Reference 1 (Figure 2). While the average fold-change increase in peak intensity due to the addition of Gd was modest (1.25-fold), we observed large fold-change increases (>2-fold) for the lowest intensity resonances (Figure 2c). Thus, the addition of Gd could improve the ability to detect low abundant metabolites. Of note, significant differences were observed in the intensity for individual metabolites, suggesting that Gd affects each metabolite to a different extent. Previous studies have suggested that a charge distribution, especially anionic metabolites, may be more affected by Gd [22,24]. We also verified that, for 0.25 mM Gd, the optimal d1 remained at 0.8 s as measured by both an increase in mean peak intensity and mean S/N (Figure S3). Taken together, our results suggest that the addition of Gd can improve both S/N and peak intensities, which will result in an overall sensitivity improvement, leading to a higher accuracy and precision in the measurement of metabolite concentrations.

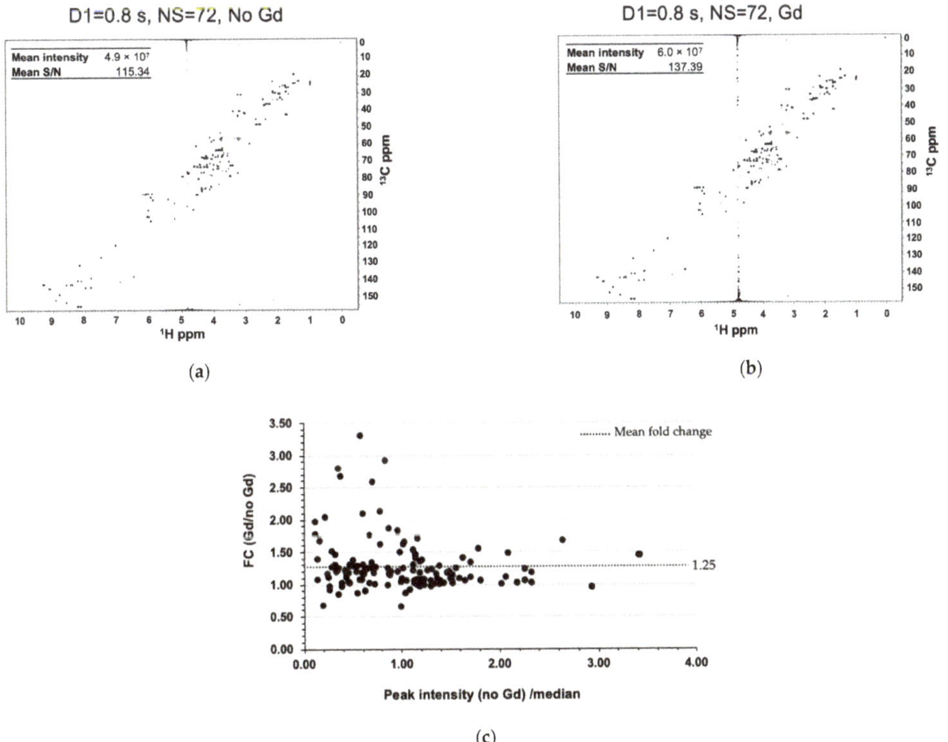

Figure 2. Gadolinium improves S/N and mean intensity of metabolite resonances. 2D ^{1}H-^{13}C HSQC spectra (**a**) without and (**b**) with the addition of Gd. (**c**) Fold change of the median normalized peak intensity in the presence of Gd.

2.3. Gadolinium Maintains Linearity

Metabolomics requires quantification across a broad range of concentrations and the ability to accurately detect changes in metabolite levels [33,34]. We previously demonstrated that NUS ^{1}H-^{13}C HSQC metabolite profiling is highly linear in the 0.05 µM to 2 mM range [15]. Rai and colleagues also observed that the addition of a PRE agent, Cu(EDTA), maintained linearity for an NUS ^{1}H-^{13}C HSQC experiment that measured four amino acids (glycine, alanine, valine and methionine) over a concentration range of 24 to 78 mM [13]. We first sought to confirm that the addition of Gd maintained linearity over a broad concentration range. A series of six NUS ^{1}H-^{13}C HSQC spectra were recorded for a mixture containing 29 metabolites (Reference 2) with concentrations ranging from 50 µM to 2 mM (Table S2). For each NMR resonance, the peak intensity was plotted as a function of concentration and the data were fit to a linear regression model (Figure S4). Example plots of the four NMR resonance peaks for leucine and the single resonance peak for pyruvic acid are shown in Figure 3. More than 98% of the metabolite resonances displayed a correlation coefficient of $R^2 > 0.9$, indicating excellent linearity (Table 1). Interestingly, glucose resonances, which can be affected by isomers and conformational changes, had an $R^2 > 0.99$ that was an improvement from our previous findings without Gd, where we observed an R^2 of ~0.8 [15]. Overall, this analysis demonstrated that peak intensities are highly linear as a function of metabolite concentration for NUS ^{1}H-^{13}C HSQC spectra in the presence of Gd.

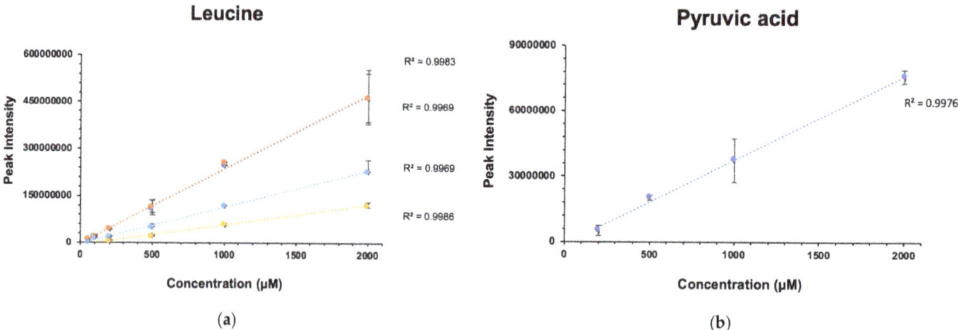

Figure 3. Linear regression analysis of NMR resonance intensities for (**a**) leucine and (**b**) pyruvic acid peaks.

Table 1. Reference 2 metabolite resonances and their correlation coefficient (R^2) for each resonance.

Metabolites/R^2	1	2	3	4	5	6	7	8	9	10	11	12	13
NAD	0.996	0.999	0.999	0.997	1.000	0.999	0.998	0.999	1.000	0.999	1.000	0.999	0.999
NADPH	1.000	1.000	0.886	0.929	0.995	0.966	0.967	0.966	0.972	0.968	0.950		
Cytidine	0.912	0.999	0.997	0.998	0.997	0.998	0.998						
UDP	0.999	0.999	0.999	0.999	0.999	0.999	1.000						
Fructose	1.000	1.000	1.000	1.000	0.984	0.996							
AMP	0.951	0.973	0.970	0.992	0.901	0.994							
Lysine	0.998	0.999	0.999	0.998	0.942								
Histidine	0.998	1.000	0.998	0.999	0.998								
Glucose	0.995	0.997	0.993	0.996	0.974								
Ribose 5P	0.983	0.990	0.989	0.996	0.994								
Glucosamine	0.997	0.959	0.992	0.951									
2-HG	0.994	0.998	0.999	0.984									
Leucine	0.998	0.997	0.999	0.997									
Nicotinic acid	0.998	0.996	0.994	0.997									
Acetylcholine	0.999	0.989	0.998	0.998									
Glutamic acid	0.998	0.998	0.952										
Malic acid	0.993	0.989	0.996										
Arginine	0.991	0.997	0.999										
Ornithine	0.998	0.996	1.000										
Choline	0.999	0.997	0.996										
Glutamine	0.998	0.998	0.998										
GTP	0.998	0.998	0.967										
Citrate	0.990	0.998											
Alanine	0.998	0.995											
Lactic acid	0.999	0.997											
Pyruvic acid	0.998												
Acetic acid	0.990												
Fumaric acid	0.989												
Succinic acid	0.998												

2.4. Gadolinium Improves the Lower Limit of Detection and Quantification

We next sought to determine the lower limit of detection (LOD) and lower limit of quantification (LOQ) for our NUS measurements in the presence of Gd. LOD and LOQ are defined as follows:

$$LOD = 3 \times \sigma \tag{1}$$

$$LOQ = 10 \times \sigma \tag{2}$$

where the variance of the noise (σ) was estimated by the median absolute deviation (*MAD*). *MAD* was calculated from the COLMAR database [35], where the positive values for all non-peak data (X_i) were used in the following equations:

$$MAD = median_i(\,|\,X_i - median_i(X_j)\,|\,) \tag{3}$$

$$\sigma = 1.4826 \times MAD \tag{4}$$

Tables 2 and 3 list the LOD and LOQ for each of the resonances detected in Reference 2. Metabolites with multiple resonances have an LOD/LOQ for each observed peak, and thus metabolites with multiple peaks will have a range of LOD/LOQ values. The average LOD and LOQ in the presence of Gd was $7.8 \pm 0.3\ \mu M$ and $26 \pm 1\ \mu M$, respectively. This is a dramatic improvement over our previous findings that yielded an average LOD and LOQ of $19.1\ \mu M$ and $65.6\ \mu M$, respectively [15]. These prior NMR experiments lacked the addition of Gd and used a longer d1 of 1.5 s. Thus, it is possible to detect lower abundant metabolites by adding Gd and decreasing d1. We also compared the effects of different NMR probes on LOD/LOQ. For the same d1 of 1.5 s, a TCI helium-cooled probe had a lower LOD/LOQ compared to a TXI nitrogen-cooled probe (Table S3).

2.5. Gadolinium Maintains Reproducibility

Highly reproducible measurements are required to detect changes in the large number of samples associated with metabolomics studies. We previously demonstrated that intensity measurements from NUS ^{1}H-^{13}C HSQC experiments with a d1 of 1.5 s were highly reproducible as evident by a percent coefficient of variation (%CV) of $14 \pm 9\%$ for a model mixture containing 15 metabolites at a concentration of $500\ \mu M$ [15]. By decreasing the d1 to 0.8 s, we observed a decrease in the %CV to $8 \pm 8\%$ (Figure 4) for three replicates of Reference 1. This was expected, given that the increased number of scans would lead to an increase in peak intensities. We only observed a modest decrease in %CV to $7 \pm 7\%$ (Figure 4) by adding Gd to the samples while maintaining a d1 of 0.8 s. This suggests that the addition of Gd does not negatively impact the reproducibility of NUS ^{1}H-^{13}C HSQC experiments and may increase the reliability of these measurements.

2.6. Gadolinium Effect on Plasma Metabolites

We next assessed the effects of Gd on our ability to detect and quantify metabolites using a commercially available standard pooled human plasma sample. We recorded a 25% NUS ^{1}H-^{13}C HSQC with or without the addition of Gd, and with a relaxation delay of 0.8 s, a constant scan number of 72, and an acquisition time of ~1 h. The addition of Gd led to a 1.12-fold increase in overall mean peak intensities. This increase was slightly less pronounced than the fold change of 1.25 observed with the model mixture and could be due to the presence of additional anions and salts, which are known to influence the impact of PRE agents [22,24]. Nonetheless, as noted for the model mixtures, we observed that the largest increase in fold change was associated with low-intensity resonances. These results further suggest that the addition of Gd could improve our ability to detect low abundant metabolites (Figure 5c). Furthermore, the % CV was lowered from 15% to 10% for the pooled human plasma sample in the presence of Gd (Figure 5d). Collectively, these results suggest that the addition of Gd to plasma samples increases the S/N for metabolite NMR resonances, especially for low abundant metabolites, and increases the reproducibility of intensity measurements. Overall, the addition of Gd to a metabolomics sample could facilitate an increase in the confidence and reliability in the detection and quantification of metabolite NMR resonances.

Table 2. Limit of detection (LOD) for measured metabolite resonances.

Metabolites														Minimal

Metabolites	1	2	3	4	5	6	7	8	9	10	11	12	13	Minimal Conc. (μM)
NAD	12.7 ± 0.49	12.58 ± 0.26	26.05 ± 0.32	12.2 ± 0.04	10.61 ± 0.19	15.54 ± 0.97	7.9 ± 0.08	13.48 ± 0.15	16.5 ± 0.04	14.41 ± 0.11	12.98 ± 0.16	9.2 ± 0.09	12.9 ± 0.02	7.90 ± 0.08
NADPH	26.85 ± 0.21	21.76 ± 0.99	7.46 ± 0.17	19.01 ± 0.38	13.4 ± 0.48	10.66 ± 0.42	14.4 ± 0.14	25.37 ± 0.5	19.46 ± 0.96	12.98 ± 0.27	24.4 ± 0.62			7.46 ± 0.17
Cytidine	18.1 ± 0.66	8.44 ± 0.02	9.04 ± 0.08	7.4 ± 0.00	6.62 ± 0.05	7.86 ± 0.06	7.89 ± 0.05							6.62 ± 0.05
UDP	6.78 ± 0.09	9.47 ± 0.15	9.82 ± 0.2	9.85 ± 0.01	8.69 ± 0.11	10.65 ± 0.15	10.69 ± 0.05							6.78 ± 0.09
Fructose	19.58 ± 0.93	13.82 ± 0.45	13.85 ± 0.44	18.89 ± 1.17	18.71 ± 0.33	9.49 ± 0.09								9.49 ± 0.09
AMP	8.35 ± 1.68	12.14 ± 1.54	3.94 ± 0.16	13.17 ± 2.97	10.75 ± 1.7	14.96 ± 2.96								3.94 ± 0.16
Lysine	23.24 ± 0.19	10.07 ± 0.24	9.28 ± 0.13	6.59 ± 0.03	3.6 ± 0.1									3.60 ± 0.10
Histidine	19.15 ± 2.19	18.79 ± 1.91	10.85 ± 1.66	7.9 ± 0.38	11.3 ± 2.09									7.90 ± 0.38
Glucose	21.62 ± 3.18	18.61 ± 2.01	16.35 ± 1.1	14.32 ± 0.11	15.79 ± 2.56									14.32 ± 0.11
Ribose 5P	41.21 ± 2.58	30.31 ± 1.82	21.22 ± 0.59	8.98 ± 0.26	19.94 ± 0.71									8.98 ± 0.26
Glucosamine	30.31 ± 1.11	19.2 ± 0.570.57	33.83 ± 1.51	22.32 ± 1.71										19.20 ± 0.57
2-HG	33.83 ± 1.76	33.3 ± 1.89	25.95 ± 0.75	12.7 ± 0.3										12.70 ± 0.3
Leucine	5.64 ± 0.98	5.58 ± 1.07	20.77 ± 1.54	11.49 ± 1.85										5.58 ± 1.07
Nicotinic acid	10.78 ± 0.03	9.71 ± 0.19	8.04 ± 0.16	7.26 ± 0.01										7.26 ± 0.01
Acetylcholine	8.56 ± 0.12	0.71 ± 0.02	8.42 ± 0.1	8.06 ± 0.07										0.71 ± 0.02
Glutamic acid	22.1 ± 1.1	8.87 ± 0.13	4.09 ± 0.08											4.09 ± 0.08
Malic acid	30.52 ± 2.05	30.12 ± 0.41	8.98 ± 0.26											8.98 ± 0.26
Arginine	25.49 ± 2.75	9.06 ± 0.09	6.09 ± 0.06											6.09 ± 0.06
Ornithine	28.27 ± 1.77	4.59 ± 0.05	6.71 ± 0.05											4.59 ± 0.05
Choline	1.41 ± 0.05	8.09 ± 0.3	7.76 ± 0.29											1.41 ± 0.05
Glutamine	8.32 ± 0.13	9.46 ± 0.22	5.06 ± 0.07											5.06 ± 0.07
GTP	11.66 ± 0.18	15.58 ± 0.51	7.54 ± 1.44											7.54 ± 1.44
Citrate	8.2 ± 1.11	8.15 ± 1.21												5.14 ± 0.01
Alanine	5.14 ± 0.01	13.4 ± 0.08												8.15 ± 1.21
Lactic acid	5.4 ± 0.02	13.89 ± 0.22												5.4 ± 0.02
Pyruvic acid	30.79 ± 2.83													30.79 ± 2.83
Acetic acid	7.21 ± 0.08													7.21 ± 0.08
Fumaric acid	5.6 ± 0.14													5.67 ± 0.12
Succinic acid	3.21 ± 0.03													3.21 ± 0.03

Table 3. Limit of quantification (LOQ) for measured metabolite resonances.

| Metabolites | LOQ (μM) per NMR Resonance | | | | | | | | | | | | | Minimal Conc. (μM) |

Metabolites	1	2	3	4	5	6	7	8	9	10	11	12	13	Minimal Conc. (μM)
NAD	42.34 ± 1.62	41.92 ± 0.88	86.83 ± 1.06	40.66 ± 0.13	35.37 ± 0.62	51.78 ± 3.22	26.35 ± 0.27	44.92 ± 0.51	55.01 ± 0.14	48.03 ± 0.38	43.26 ± 0.55	30.66 ± 0.3	43.01 ± 0.06	26.35 ± 0.27
NADPH	89.49 ± 0.7	72.53 ± 3.29	24.87 ± 0.55	63.38 ± 1.25	44.66 ± 1.61	35.52 ± 1.39	48.01 ± 0.47	84.56 ± 1.67	64.88 ± 3.19	43.27 ± 0.91	81.33 ± 2.07			35.52 ± 1.39
Cytidine	60.34 ± 2.2	28.12 ± 0.08	30.15 ± 0.28	24.68 ± 0.01	22.05 ± 0.18	26.2 ± 0.19	26.29 ± 0.15							22.05 ± 0.18
UDP	22.59 ± 0.29	31.55 ±0.51	32.74 ± 0.68	32.85 ± 0.03	28.95 ± 0.36	35.49 ± 0.49	35.64 ± 0.15							22.59 ± 0.29
Fructose	65.27 ± 3.1	46.06 ±1.5	46.18 ± 1.47	62.98 ± 3.89	62.38 ± 1.09	31.63 ± 0.29								31.63 ± 0.29
AMP	27.83 ± 5.61	40.46 ± 5.12	13.13 ± 0.52	43.92 ± 9.89	35.84 ± 5.68	49.86 ± 9.88								13.13 ± 0.52
Lysine	77.47 ± 0.64	33.57 ± 0.8	30.92 ± 0.44	21.95 ± 0.08	12.01 ± 0.34									12.01 ± 0.34
Histidine	63.83 ± 7.29	62.63 ± 6.38	36.18 ± 5.53	26.32 ± 1.25	37.67 ± 6.97									26.32 ± 1.25
Glucose	72.07 ±10.59	62.02 ± 6.69	54.5 ± 3.66	47.74 ± 0.36	52.63 ± 8.54									47.74 ± 0.36
Ribose 5P	137.37 ± 8.59	101.02 ± 6.08	70.73 ± 1.97	30.11 ± 0.89	66.45 ± 2.38									30.11 ± 0.89
Glucosamine	101.02 ± 3.71	42.87 ± 2.83	112.76 ± 5.02	74.4 ± 5.69										42.87 ± 2.83
2-HG	112.76 ± 5.86	110.99 ± 6.3	86.49 ± 2.51	42.34 ±1.00										42.34 ± 1.00
Leucine	18.78 ± 3.27	18.61 ± 3.58	69.24 ± 5.13	38.28 ± 6.16										18.61 ± 3.58
Nicotinic acid	35.93 ± 0.1	32.36 ± 0.62	26.8 ± 0.54	24.19 ± 0.04										24.19 ± 0.04
Acetylcholine	28.54 ± 0.4	2.38 ± 0.08	28.08 + BL86 ± 0.33	26.88 ± 0.24										2.38 ± 0.08
Glutamic acid	73.67 ± 3.65	29.55 ± 0.44	13.64 ± 0.28											13.64 ± 0.28
Malic acid	101.74 ± 6.84	100.41 ± 1.36	29.92 ± 0.88											29.92 ± 0.88
Arginine	84.96 ± 9.16	30.2 ± 0.3	20.32 ± 0.19											20.32 ± 0.19
Ornithine	94.24 ± 5.91	15.3 ± 0.16	22.35 ± 0.18											15.30 ± 0.16
Choline	4.69 ± 0.15	26.98 ± 0.99	25.85 ± 0.97											4.69 ± 0.15
Glutamine	27.72 ± 0.44	31.54 ± 0.73	16.85 ± 0.24											16.85 ± 0.24
GTP	38.87 ± 0.6	51.94 ± 1.7	25.13 ± 4.8											25.13 ± 4.80
Citrate	27.33 ± 3.71	27.18 ± 4.03												27.18 ± 4.03
Alanine	17.15 ± 0.02	44.67 ± 0.26												17.15 ± 0.02
Lactic acid	18.08 ± 0.19	46.29 ± 0.74												18.08 ± 0.19
Pyruvic acid	102.65 ± 9.42													102.65 ± 9.42
Acetic acid	24.61 ± 0.35													24.61 ± 0.35
Fumaric acid	18.89 ± 0.39													18.89 ± 0.39
Succinic acid	10.69 ± 0.11													10.69 ± 0.11

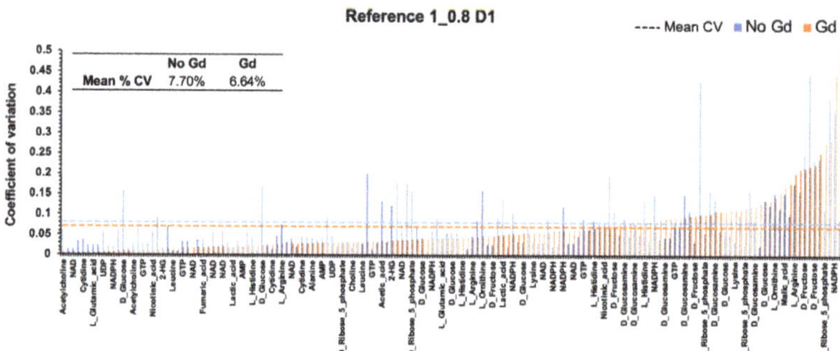

Figure 4. Gadolinium maintains the reproducibility of NMR intensity measurements. The percent coefficient of variation (%CV) in the peak intensities measured from NUS ^{1}H-^{13}C HSQC spectra. The %CV decreased for metabolite resonances in the presence of Gd (orange) compared to no Gd (blue).

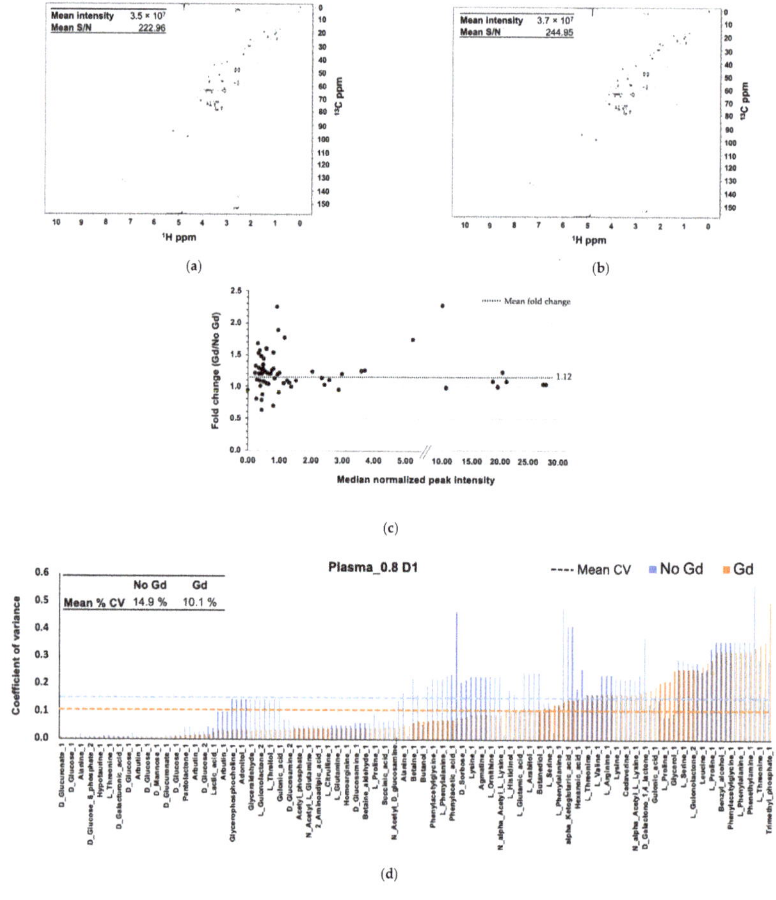

Figure 5. Gadolinium improves S/N and the mean intensity of plasma metabolite resonances. ^{1}H-^{13}C HSQC spectra (**a**) without and (**b**) with the addition of gadolinium. (**c**) Fold change of median normalized intensity in the presence of gadolinium. (**d**) Percent coefficient of variation (% CV) measured intensity decreases for metabolite resonances in the presence of Gd (orange) compared to no Gd (blue).

3. Discussion

Metabolites are influenced both by the genome and the environment, and thus provide the most comprehensive readout for the state of an individual [36–38]. By monitoring changes in metabolites, it is possible to develop novel biomarkers that reveal important health information. Indeed, altered metabolite levels have been observed in many diseases, including diabetes [39], neurodegeneration [40], cancer [4], cardiovascular disease [6], and even aging [3]. Furthermore, in a series of separate studies, we have identified metabolite biomarkers of response (BoRs) that correlate with drug responsiveness for metastatic breast cancer patients treated with CDK4/6 inhibitors as well as the anti-HER2 therapy trastuzumab; and for gastrointestinal stromal tumor (GIST) patients treated with tyrosine kinase inhibitors [41–43]. While additional validation studies are required, these preliminary results suggest the exciting possibility that metabolite-based biomarkers have for designing optimal treatment strategies for individual patients, which is a major goal of precision medicine.

To uncover metabolite BoRs, it is first necessary to accurately measure metabolite levels in biospecimens collected from a large number of patients so that the relative metabolite concentration can be correlated with disease outcomes and/or a drug response. NMR and mass spectrometry (MS) are the two most commonly used analytical platforms for measuring metabolites. Traditionally, MS has been favored due to its high sensitivity, dynamic range, and potential for high throughput. There are numerous sensitive LC-MS methods reported in the literature for the identification of endogenous metabolites in human plasma [19–21,44–46]. For example, amino acids are routinely detected at submicromolar concentrations (0.01 to 0.04 μM) by these targeted LC-MS methods. In contrast, NMR-based approaches typically detect plasma concentrations in the micromolar (3–10 μM) range [44–46]. However, MS can suffer from reproducibility issues, requires chromatography because of the narrow molecular-weight distribution of metabolites, and still faces challenges in metabolite identification [9]. Conversely, NMR is highly reproducible and can reveal structural information to facilitate metabolite identification. However, NMR is limited by sensitivity and spectral overlap [47]. Multidimensional NMR can overcome some of these challenges but requires extremely long experimental times that are not practical for the large number of samples needed for BoR discovery. Efforts to increase the throughput of NMR are actively being explored. We and others have demonstrated that NUS can accelerate NMR acquisition times to meet the high-throughput demands of metabolomics [13–15]. In an approximate one-hour experiment, we verified that intensity measurements from NUS ^{1}H-^{13}C HSQC spectra are highly reproducible and can facilitate the detection of a wide variety of metabolites in the low micromolar range.

In this study, we sought to explore additional techniques to extend the limit of metabolite detection by multidimensional NMR. As a first step, we assessed the effect of the d1 relaxation delay on S/N. The relaxation delay is the experimental time between scans in an NMR experiment to allow the nuclear spins to return to equilibrium, which is influenced by T_1 longitudinal relaxation rates of each nuclei in the sample. For quantitative NMR, it is suggested to set d1 to at least 5 times the slowest T_1 [17]. For metabolomics, this is not practical as T_1s can be several seconds in length or longer. Instead, d1 is commonly set to a shorter, predetermined value for semi-quantification. Herein, we demonstrated that a decrease in d1 from 1.5 s to 0.8 s enabled an increase in the number of scans from 36 to 72, which led to an overall improvement in S/N and an increase in the mean signal intensity for metabolite resonances. Notably, this was accomplished without increasing the total time to acquire the NMR spectrum. Any further reduction in d1 was observed to result in severe signal artifacts from the solvent.

With the optimal d1 selected, we next sought to manipulate the T_1s of metabolite nuclei through PRE. PRE accelerates spin relaxation due to induced magnetic dipolar interactions with unpaired electrons. PRE-based applications have been used for macromolecular structure determination, characterizing long-range interactions and identifying transiently populated states of proteins and complexes [28,48,49]. PRE agents also provide the foundation for contrast agents in magnetic resonance imaging (MRI) [50]. Previous metabolomics studies have suggested that the addition of PRE-agents, Cu(EDTA) or Gd, can decrease T_1 relaxation times for metabolites [13,17]. We observed similar trends using our standard 25% NUS ^{1}H-^{13}C HSQC experimental parameters and noted that the addition of Gd led to an overall improvement in S/N and mean signal intensity for metabolites in a model mixture and from plasma samples. Although the average increased fold change in intensity with Gd was relatively modest, we did observe a significant improvement (>2-fold increase) for metabolite resonances with the lowest signal intensity. For metabolomics studies, the ability to accurately detect and quantify a broad range of metabolites that span different chemical classes and concentration ranges is paramount. Thus, an increase in the NMR signal intensity for low abundant metabolites suggests that the addition of Gd could improve the coverage of the metabolome. Indeed, both the lower limit of detection and quantification (LoD/LoQ) were significantly improved in the presence of Gd. In our previous results, the LoD and LoQ for a model mixture of metabolites was 19.1 µM and 65.6 µM, respectively. For the same model mixture, the LoD and LoQ decreased by more than 2-fold to 7.8 µM and 26 µM, respectively, by decreasing the d1 and by the addition of Gd.

4. Materials and Methods

Commercially available analytical standards were used to prepare model mixtures of metabolites, Reference 1 and Reference 2 (Tables S1 and S2): acetylcholine chloride ($C_7H_{15}NO_2 \cdot HCl$, >99%), L-arginine ($C_6H_{14}N_4O_2$, >98%), L-glutamine ($C_5H_{10}N_2O_3$, >99%), D-alpha-hydroxyglutaric acid disodium salt ($C_5H_6Na_2O_5$, >98%), α-ketoglutaric acid disodium salt dihydrate ($C_5H_4Na_2O_5 \cdot 2H_2O$, >98%), adenosine 5-monophosphate disodium ($C_{10}H_{12}N_5Na_2O_7P$, >99%), D-(-)-fructose ($C_6H_{12}O_6$, >99%), guanosine 5-triphosphate sodium salt ($C_{10}H_{16}N_5O_{14}P_3 \cdot xNa + yH_2O$, >95%), lithium potassium acetyl phosphate ($C_2H_3KLiO_5P$, >97%), L-ornithine hydrochloride ($C_5H_{12}N_2O_2 \cdot HCl$, >98%), β-nicotinamide adenine dinucleotide hydrate ($C_{21}H_{27}N_7O_{14}P_2 \cdot xH_2O$, >98%), DL-malic acid ($C_4H_6O_5$, >99%), D-ribose 5-phosphate disodium salt dihydrate ($C_5H_9Na_2O_8P \cdot 2H_2O$, >99%), sodium succinate dibasic hexahydrate ($C_4H_4Na_2O_4 \cdot 6H_2O$, >99%), sodium acetate ($C_2H_3NaO_2$, >99%), sodium L-lactate ($C_3H_5NaO_3$, >99), sodium citrate tribasic dihydrate ($C_6H_5O_7Na_3 \cdot 2H_2O$, >99%), sodium fumarate dibasic ($C_4H_2Na_2O_4$, >98%), sodium pyruvate ($C_3H_3NaO_3$, >99%), uridine 5-diphosphate ($C_9H_{12}N_2Na_2O_{12}P_2 \cdot xH_2O$, >96%), L-alanine ($C_3H_7NO_2$, >98%), L-cysteine ($C_3H_7NO_2S$, >98%), D-(+)-glucosamine hydrochloride ($C_6H_{13}NO_5 \cdot HCl$, >99%), D-(+)-glucose ($C_6H_{12}O_6$, >99.5%), choline chloride ($C_5H_{13}NO \cdot HCl$, >99%), cytidine ($C_9H_{13}N_3O_5$, >99%), L-leucine ($C_6H_{13}NO_2$, >98.5%), L-glutamic acid monosodium salt monohydrate ($C_5H_8NNaO_4 \cdot H_2O$, >99%), L-histidine ($C_6H_{11}N_3O_3 \cdot HCl$, >98.5%), L-lysine, monohydrochloride ($C_6H_{14}N_2O_2 \cdot HCl$, >98.5%). All the compounds were obtained from Sigma-Aldrich. Deuterium oxide (D_2O, 99.0%) was purchased from Cambridge Isotope Laboratory, Inc., Andover, MA. Pooled human plasma (apheresisderived, K2EDTA) was purchased from innovative research, Novi, MI. Paramagnetic relaxation agents gadobutrol (Gd) ($C_{18}H_{31}GdN_4O_9$, >99.9%) and copper (II) disodium ethylenediaminetetraacetate tetrahydrate (Cu- EDTA) ($C_{10}H_{12}CuN_2Na_2O_8 \cdot 4H_2O$) were procured from MedChemExpress, Monmouth Junction, NJ and TCI America, Portland, OR respectively. The NMR reference standard, deuterated 3-(trimethylsilyl)-1-propanesulfonic acid sodium salt (DSS-d6, 98%) was purchased from Cambridge Isotope Laboratory, Andover, MA, USA.

4.1. NMR Sample Preparation

Reference 1 and Reference 2 were prepared as previously described [15]. Human plasma extraction: metabolites were extracted from 1 mL of human plasma via a methanol and chloroform liquid–liquid extraction. The aqueous phase was transferred to a 15 mL Falcon tube and freeze-dried. The powder was reconstituted in 180 µL of 50 mM phosphate buffer at pH 7.4 in D_2O, and then immediately transferred to a 3 mm NMR tube for NMR data collection. The NMR standard, DSS-d6, was added to each sample for chemical shift referencing.

4.2. NMR Experiments and Data Processing

All NMR spectra were acquired on a Bruker AVANCE III solution-state NMR spectrometer equipped with a liquid helium-cooled TCI (H/F, C, N), deuterium lock, and a cryoprobe operating at a frequency of 599.773010 MHz for proton and 150.822998 MHz for carbon. NUS schedules were generated using a Poisson gap distribution with a sinusoidal weight of two and random seed generator [51]. The same 25% NUS schedule and seed were used for all experiments. All NMR data were collected at 298 K.

The spectral widths along the direct and the indirect dimensions were set at 9578.544 and 24,132.982 Hz, respectively. The number of complex points in the direct dimension was set at 512, and in the indirect dimension set at 32 with a 25% NUS sampling schedule. The number of scans for the 1D ^{1}H experiments was set to 64 (d1 = 1.5 s), 72 (d1 = 1.2 s), 84 (d1 = 0.8 s), and 92 (d1 = 0.6 s). The number of scans for the 2D ^{1}H-^{13}C HSQC experiments was set to 36 (d1 = 1.5 s), 48 (d1 = 1.2 s), 72 (d1 = 0.8 s) and 84 (d1 = 0.6 s), respectively. The scan numbers were selected such that the total acquisition time for each 1D- and 2D-experiment was on average 246 s, and 69 min., respectively. The transmitter frequency offset was set to 75 ppm in the ^{13}C dimension and 4.7 ppm in the ^{1}H dimension.

The spectral data were processed using the NMRPipe software package, as previously described [52]. The NUS data were reconstructed using iterative soft thresholding according to the hmsIST algorithm [51] to generate the same number of direct dimension data points and twice the number of indirect dimension data points, 512 (N_2) × 256 (N_1). Both the NUS and US NMR data were zero-filled, Fourier-transformed and manually phase-corrected to yield a final digital resolution of 2048 (N_2) × 2048 (N_1) points. Chemical shift queries, metabolite identifications and quantifications were performed using the COLMARm NMR webserver (http://spin.ccic.ohio-state.edu/index.php/colmar (accessed on 4 February 2021) [35]. The metabolite list is presented in the Supplementary Materials section for Reference 1 (Table S1) and Reference 2 (Table S2). The resonance assignments were used as previously reported [15].

5. Conclusions

In this study, we demonstrated the ability of Gd to improve the sensitivity of 2D NUS NMR spectra for the analysis of metabolomics samples. The addition of Gd led to an overall improvement in S/N and mean signal intensity for metabolites in both a model mixture and plasma samples. In the model mixture, the addition of Gd led to a 1.25-fold improvement in NMR signal intensities, which resulted in 1.7- and 1.6-fold improvements in LOD and LOQ, respectively. Interestingly, a significant improvement (>2-fold increase) was observed for metabolites with the lowest peak intensities, which suggests that the combination of Gd with NUS may improve the coverage of the plasma metabolome. The addition of Gd also maintained the highly linear intensity measurements that were correlated with a wide range of metabolite concentrations (50 µM to 2 mM). The reproducibility of intensity measurements, as noted by a decrease in %CV for both the model mixture (8% to 7%) and the plasma samples (15% to 10%), was similarly improved with the addition of Gd. Collectively, our results suggest that supplementing metabolomics samples with 0.25 mM Gd can improve the sensitivity of 2D NUS ^{1}H-^{13}C HSQC spectra and enhance the overall quality of the resulting data analysis. The routine adoption of PRE by the metabolomics community may expand the utility of multidimensional NMR to empower future biomarker discoveries.

Supplementary Materials: The following are available online. Figure S1. Chemical structure of Gadobutrol (Gd), Figure S2. Effect of addition of relaxation agents on 1H-13C HSQC 1D ^{1}H spectral intensity for a model mixture (Reference 1) of metabolites in time equivalent experiments, Figure S3. Plots showing (a) mean peak intensity and (b) mean S/N for Reference 1 with and without addition of Gadobutrol, Figure S4. Linear regression curve of Reference 2 metabolites, Table S1. The list of metabolites in Reference 1 model mixture, Table S2. The list of metabolites in Reference 2 model mixture, Table S3. Comparison of LOD and LOQ at 0.8 D1 and at 1.5 D1 with and without Gd.

Author Contributions: Conceptualization, R.P. and E.M.O.; investigation, C.H., N.T. and B.Z.; data curation, C.H.; writing—original draft preparation, C.H. and E.M.O.; writing—review and editing, R.P. and E.M.O. All authors have read and agreed to the published version of the manuscript.

Funding: This research was funded by Olaris, Inc., and, in part, by funding from the National Science Foundation under Grant Number (1660921) to R.P. and the National Institutes of Health Grant Number (P20 GM113126, NIGMS, Nebraska Center for Integrated Biomolecular Communication) to R.P. Any opinions, findings, and conclusions or recommendations expressed in this material are those of the author(s) and do not necessarily reflect the views of the National Science Foundation.

Institutional Review Board Statement: Not Applicable.

Informed Consent Statement: Not Applicable.

Data Availability Statement: Data available on request due to proprietary restrictions. The data presented in this study are available on request from the corresponding author.

Acknowledgments: The authors would like to thank Leonardo O. Rodrigues, Chen Dong, Srihari Raghavendra Rao, Tan Chui-Mae, Josephine J. Wolf, Kanchan Sonkar, Monty Mahantesh Kothiwale, Charles Sang, Gerhard Wagner, Amber Balazs and David Emmons, for their many valuable discussions that motivated the writing of this communication.

Conflicts of Interest: The authors report that this work (design of the study, collection, analysis, and interpretation of data) was supported by Olaris, Inc. Chandrashekhar Honrao, Nathalie Teissier and Elizabeth O'Day are employees at Olaris, Inc. Bo Zhang is a former employee. Robert Powers is a member of the Scientific Advisory Board at Olaris, Inc.

Sample Availability: Samples are or not available from the authors.

References

1. Riekeberg, E.; Powers, R. New frontiers in metabolomics: From measurement to insight. *F1000Research* **2017**, *6*, 1148. [CrossRef]
2. Li, L.; Krznar, P.; Erban, A.; Agazzi, A.; Martin-Levilain, J.; Supale, S.; Kopka, J.; Zamboni, N.; Maechler, P. Metabolomics Identifies a Biomarker Revealing In Vivo Loss of Functional β-Cell Mass Before Diabetes Onset. *Diabetes* **2019**, *68*, 2272–2286. [CrossRef]
3. Shao, Y.; Le, W. Recent advances and perspectives of metabolomics-based investigations in Parkinson's disease. *Mol. Neurodegener.* **2019**, *14*, 1–12. [CrossRef]
4. Puchades-Carrasco, L. Metabolomics Applications in Precision Medicine: An Oncological Perspective. *Curr. Top. Med. Chem.* **2017**, *17*, 2740–2751. [CrossRef]
5. Cirulli, E.T.; Guo, L.; Swisher, C.L.; Shah, N.; Huang, L.; Napier, L.A.; Kirkness, E.F.; Spector, T.D.; Caskey, C.T.; Thorens, B.; et al. Profound Perturbation of the Metabolome in Obesity Is Associated with Health Risk. *Cell Metab.* **2019**, *29*, 488–500. [CrossRef]
6. Ussher, J.R.; Elmariah, S.; Gerszten, R.E.; Dyck, J.R. The Emerging Role of Metabolomics in the Diagnosis and Prognosis of Cardiovascular Disease. *J. Am. Coll. Cardiol.* **2016**, *68*, 2850–2870. [CrossRef] [PubMed]
7. Darst, B.F.; Koscik, R.L.; Hogan, K.J.; Johnson, S.C.; Engelman, C.D. Longitudinal plasma metabolomics of aging and sex. *Aging* **2019**, *11*, 1262–1282. [CrossRef] [PubMed]
8. Bhinderwala, F.; Powers, R. NMR Metabolomics Protocols for Drug Discovery. In *Cardiovascular Development*; Springer Science and Business Media LLC: Berlin, Germany, 2019; Volume 2037, pp. 265–311.
9. Markley, J.L.; Brüschweiler, R.; Edison, A.; Eghbalnia, H.R.; Powers, R.; Raftery, D.; Wishart, D.S. The future of NMR-based metabolomics. *Curr. Opin. Biotechnol.* **2017**, *43*, 34–40. [CrossRef] [PubMed]
10. McAlpine, J.B.; Chen, S.-N.; Kutateladze, A.; MacMillan, J.B.; Appendino, G.; Barison, A.; Beniddir, M.A.; Biavatti, M.W.; Bluml, S.; Boufridi, A.; et al. The value of universally available raw NMR data for transparency, reproducibility, and integrity in natural product research. *Nat. Prod. Rep.* **2019**, *36*, 35–107. [CrossRef]
11. Dona, A.; Kyriakides, M.; Scott, F.; Shephard, E.; Varshavi, D.; Veselkov, K.; Everett, J.R. A guide to the identification of metabolites in NMR-based metabonomics/metabolomics experiments. *Comput. Struct. Biotechnol. J.* **2016**, *14*, 135–153. [CrossRef] [PubMed]
12. Bingol, K.; Bruschweiler-Li, L.; Li, D.-W.; Zhang, B.; Xie, M.; Brüschweiler, R. Emerging new strategies for successful metabolite identification in metabolomics. *Bioanalysis* **2016**, *8*, 557–573. [CrossRef]
13. Rai, R.; Sinha, N. Fast and Accurate Quantitative Metabolic Profiling of Body Fluids by Nonlinear Sampling of1H–13C Two-Dimensional Nuclear Magnetic Resonance Spectroscopy. *Anal. Chem.* **2012**, *84*, 10005–10011. [CrossRef]
14. Von Schlippenbach, T.; Oefner, P.J.; Gronwald, W. Systematic Evaluation of Non-Uniform Sampling Parameters in the Targeted Analysis of Urine Metabolites by 1H,1H 2D NMR Spectroscopy. *Sci. Rep.* **2018**, *8*, 1–10. [CrossRef]
15. Zhang, B.; Powers, R.; O'Day, E.M. Evaluation of Non-Uniform Sampling 2D 1H–13C HSQC Spectra for Semi-Quantitative Metabolomics. *Metabolites* **2020**, *10*, 203. [CrossRef] [PubMed]

16. Sørensen, O.W. James Keeler. Understanding NMR Spectroscopy. *Magn. Reson. Chem.* **2006**, *44*, 820. [CrossRef]
17. Mulder, F.A.A.; Tenori, L.; Luchinat, C. Fast and Quantitative NMR Metabolite Analysis Afforded by a Paramagnetic Co-Solute. *Angew. Chem. Int. Ed.* **2019**, *58*, 15283–15286. [CrossRef]
18. Cai, X.; Li, R. Concurrent profiling of polar metabolites and lipids in human plasma using HILIC-FTMS. *Sci. Rep.* **2016**, *6*, 36490. [CrossRef]
19. Simón-Manso, Y.; Lowenthal, M.S.; Kilpatrick, L.E.; Sampson, M.L.; Telu, K.H.; Rudnick, P.A.; Mallard, W.G.; Bearden, D.W.; Schock, T.; Tchekhovskoi, D.V.; et al. Metabolite Profiling of a NIST Standard Reference Material for Human Plasma (SRM 1950): GC-MS, LC-MS, NMR, and Clinical Laboratory Analyses, Libraries, and Web-Based Resources. *Anal. Chem.* **2013**, *85*, 11725–11731. [CrossRef]
20. Dunn, W.B.; Broadhurst, D.; Begley, P.; Zelena, E.; Francis-McIntyre, S.; Anderson, N.; Brown, M.; Knowles, J.D.; Halsall, A.; Haselden, J.N.; et al. Procedures for large-scale metabolic profiling of serum and plasma using gas chromatography and liquid chromatography coupled to mass spectrometry. *Nat. Protoc.* **2011**, *6*, 1060–1083. [CrossRef] [PubMed]
21. Telu, K.H.; Yan, X.; Wallace, W.E.; Stein, S.E.; Simón-Manso, Y. Analysis of human plasma metabolites across different liquid chromatography/mass spectrometry platforms: Cross-platform transferable chemical signatures. *Rapid Commun. Mass Spectrom.* **2016**, *30*, 581–593. [CrossRef] [PubMed]
22. Murphy, P.S.; Leach, M.; Rowland, I.J. Signal modulation in1H magnetic resonance spectroscopy using contrast agents: Proton relaxivities of choline, creatine, andN-acetylaspartate. *Magn. Reson. Med.* **1999**, *42*, 1155–1158. [CrossRef]
23. Valkovič, L.; Lau, J.Y.C.; Abdesselam, I.; Rider, O.J.; Frollo, I.; Tyler, D.J.; Rodgers, C.T.; Miller, J.J.J. Effects of contrast agents on relaxation properties of 31 P metabolites. *Magn. Reson. Med.* **2021**, *85*, 1805–1813. [CrossRef]
24. Botta, M.; Aime, S.; Barge, A.; Bobba, G.; Dickins, R.S.; Parker, D.; Terreno, E. Ternary Complexes between Cationic GdIII Chelates and Anionic Metabolites in Aqueous Solution: An NMR Relaxometric Study. *Chem. A Eur. J.* **2003**, *9*, 2102–2109. [CrossRef] [PubMed]
25. Scott, L.J. Gadobutrol: A Review in Contrast-Enhanced MRI and MRA. *Clin. Drug Investig.* **2018**, *38*, 773–784. [CrossRef]
26. Clore, G.M.; Iwahara, J. Theory, Practice, and Applications of Paramagnetic Relaxation Enhancement for the Characterization of Transient Low-Population States of Biological Macromolecules and Their Complexes. *Chem. Rev.* **2009**, *109*, 4108–4139. [CrossRef] [PubMed]
27. Giannoulis, A.; Ben-Ishay, Y.; Goldfarb, D. *Characteristics of Gd(III) Spin Labels for the Study of Protein Conformations*; Elsevier BV: Amsterdam, The Netherlands, 2021; Volume 651, pp. 235–290.
28. Sakol, N.; Egawa, A.; Fujiwara, T. Gadolinium Complexes as Contrast Agent for Cellular NMR Spectroscopy. *Int. J. Mol. Sci.* **2020**, *21*, 4042. [CrossRef]
29. Tang, H.; Wang, Y.; Nicholson, J.; Lindon, J. Use of relaxation-edited one-dimensional and two dimensional nuclear magnetic resonance spectroscopy to improve detection of small metabolites in blood plasma. *Anal. Biochem.* **2004**, *325*, 260–272. [CrossRef] [PubMed]
30. Shahbazi-Gahrouei, D.; Williams, M.; Allen, B.J. In vitro study of relationship between signal intensity and gadolinium-DTPA concentration at high magnetic field strength. *Australas. Radiol.* **2001**, *45*, 298–304. [CrossRef] [PubMed]
31. Lee, M.-J.; Kim, M.-J.; Yoon, C.-S.; Song, S.Y.; Park, K.; Kim, W.S. The T2-Shortening Effect of Gadolinium and the Optimal Conditions for Maximizing the CNR for Evaluating the Biliary System: A Phantom Study. *Korean J. Radiol.* **2011**, *12*, 358–364. [CrossRef]
32. Rovnyak, D.; Hoch, J.; Stern, A.; Wagner, G. Resolution and sensitivity of high field nuclear magnetic resonance spectroscopy. *J. Biomol. NMR* **2004**, *30*, 1–10. [CrossRef] [PubMed]
33. Patterson, R.E.; Ducrocq, A.J.; McDougall, D.J.; Garrett, T.J.; Yost, R.A. Comparison of blood plasma sample preparation methods for combined LC–MS lipidomics and metabolomics. *J. Chromatogr. B* **2015**, *1002*, 260–266. [CrossRef] [PubMed]
34. Krug, S.; Kastenmüller, G.; Stückler, F.; Rist, M.J.; Skurk, T.; Sailer, M.; Raffler, J.; Römisch-Margl, W.; Adamski, J.; Prehn, C.; et al. The dynamic range of the human metabolome revealed by challenges. *FASEB J.* **2012**, *26*, 2607–2619. [CrossRef] [PubMed]
35. Bingol, K.; Li, D.-W.; Zhang, B.; Brüschweiler, R. Comprehensive Metabolite Identification Strategy Using Multiple Two-Dimensional NMR Spectra of a Complex Mixture Implemented in the COLMARm Web Server. *Anal. Chem.* **2016**, *88*, 12411–12418. [CrossRef]
36. Johnson, C.; Ivanisevic, J.; Siuzdak, G. Metabolomics: Beyond biomarkers and towards mechanisms. *Nat. Rev. Mol. Cell Biol.* **2016**, *17*, 451–459. [CrossRef] [PubMed]
37. Clish, C.B. Metabolomics: An emerging but powerful tool for precision medicine. *Mol. Case Stud.* **2015**, *1*, a000588. [CrossRef]
38. Fiehn, O. Metabolomics—The link between genotypes and phenotypes. *Plant Mol. Biol.* **2002**, *48*, 155–171. [CrossRef] [PubMed]
39. Hameed, I.; Masoodi, S.R.; Mir, S.A.; Nabi, M.; Ghazanfar, K.; A Ganai, B. Type 2 diabetes mellitus: From a metabolic disorder to an inflammatory condition. *World J. Diabetes* **2015**, *6*, 598–612. [CrossRef]
40. Procaccini, C.; Santopaolo, M.; Faicchia, D.; Colamatteo, A.; Formisano, L.; de Candia, P.; Galgani, M.; De Rosa, V.; Matarese, G. Role of metabolism in neurodegenerative disorders. *Metab. Clin. Exp.* **2016**, *65*, 1376–1390. [CrossRef]
41. Zhang, B.; Warner, J.; Pinto, C.; Juric, D.; ODay, E. NMR-metabolite-resonance signature to predict HR+ breast cancer patient response to CDK4/6 inhibitors. *J. Clin. Oncol.* **2019**, *37*. [CrossRef]
42. O'Day, E.; Leitzel, K.; Ali, S.M.; Zhang, B.; Dong, C.; Gu, H.; Shi, X.; Drabick, J.J.; Cream, L.; Vasekar, M.; et al. Abstract P4-10-25: Pretreatment serum metabolome predicts PFS in first-line trastuzumab-treated metastatic breast cancer. *Poster Sess. Abstr.* **2020**, *80*, 4–10.
43. Honrao, C.; Rao, S.R.; Teissier, N.; Call, S.G.; Oday, E.M.; Janku, F. Abstract LB031: Plasma based metabolic profiling in metastatic gastrointestinal stromal tumors (GIST). *Clin. Res.* **2021**, *81*, LB031. [CrossRef]

44. Gong, L.-L.; Yang, S.; Zhang, W.; Han, F.-F.; Xuan, L.-L.; Lv, Y.-L.; Liu, H.; Liu, L.-H. Targeted Metabolomics for Plasma Amino Acids and Carnitines in Patients with Metabolic Syndrome Using HPLC-MS/MS. *Dis. Markers* **2020**, *2020*, 1–8. [CrossRef] [PubMed]
45. Tomita, R.; Todoroki, K.; Maruoka, H.; Yoshida, H.; Fujioka, T.; Nakashima, M.; Yamaguchi, M.; Nohta, H. Amino Acid Metabolomics Using LC-MS/MS: Assessment of Cancer-Cell Resistance in a Simulated Tumor Microenvironment. *Anal. Sci.* **2016**, *32*, 893–900. [CrossRef]
46. Ni, J.; Xu, L.; Li, W.; Wu, L. Simultaneous determination of thirteen kinds of amino acid and eight kinds of acylcarnitine in human serum by LC-MS/MS and its application to measure the serum concentration of lung cancer patients. *Biomed. Chromatogr.* **2016**, *30*, 1796–1806. [CrossRef] [PubMed]
47. Schätzlein, M.P.; Becker, J.; Schulze-Sünninghausen, D.; Pineda-Lucena, A.; Herance, J.R.; Luy, B. Rapid two-dimensional ALSOFAST-HSQC experiment for metabolomics and fluxomics studies: Application to a 13C-enriched cancer cell model treated with gold nanoparticles. *Anal. Bioanal. Chem.* **2018**, *410*, 2793–2804. [CrossRef]
48. Clore, G.M.; Tang, C.; Iwahara, J. Elucidating transient macromolecular interactions using paramagnetic relaxation enhancement. *Curr. Opin. Struct. Biol.* **2007**, *17*, 603–616. [CrossRef] [PubMed]
49. Clore, G.M. Practical Aspects of Paramagnetic Relaxation Enhancement in Biological Macromolecules. In *Biofilms*; Elsevier BV: Amsterdam, The Netherlands, 2015; Volume 564, pp. 485–497.
50. Xiao, Y.-D.; Paudel, R.; Liu, J.; Ma, C.; Zhang, Z.-S.; Zhou, S.-K. MRI contrast agents: Classification and application (Review). *Int. J. Mol. Med.* **2016**, *38*, 1319–1326. [CrossRef]
51. Hyberts, S.G.; Milbradt, A.G.; Wagner, A.B.; Arthanari, H.; Wagner, G. Application of iterative soft thresholding for fast reconstruction of NMR data non-uniformly sampled with multidimensional Poisson Gap scheduling. *J. Biomol. NMR* **2012**, *52*, 315–327. [CrossRef]
52. Delaglio, F.; Grzesiek, S.; Vuister, G.W.; Zhu, G.; Pfeifer, J.; Bax, A. NMRPipe: A multidimensional spectral processing system based on UNIX pipes. *J. Biomol. NMR* **1995**, *6*, 277–293. [CrossRef]

molecules

Article

Establishing a Metabolite Extraction Method to Study the Metabolome of *Blastocystis* Using NMR

Jamie M. Newton [1,2], Emma L. Betts [1], Lyto Yiangou [1], Jose Ortega Roldan [2], Anastasios D. Tsaousis [1,*] and Gary S. Thompson [2,*]

[1] Laboratory of Molecular & Evolutionary Parasitology, RAPID Group, School of Biosciences, University of Kent, Canterbury CT2 7NJ, UK; jmn24@kent.ac.uk (J.M.N.); elb48@kent.ac.uk (E.L.B.); lytoyiangou@hotmail.com (L.Y.)
[2] Wellcome Trust Biomolecular NMR Facility, School of Biosciences, University of Kent, Canterbury CT2 7NJ, UK; J.L.Ortega-Roldan@kent.ac.uk
* Correspondence: A.Tsaousis@kent.ac.uk (A.D.T.); G.S.Thompson@kent.ac.uk (G.S.T.)

Abstract: *Blastocystis* is an opportunistic parasite commonly found in the intestines of humans and other animals. Despite its high prevalence, knowledge regarding *Blastocystis* biology within and outside the host is limited. Analysis of the metabolites produced by this anaerobe could provide insights that can help map its metabolism and determine its role in both health and disease. Due to its controversial pathogenicity, these metabolites could define its deterministic role in microbiome's "health" and/or subsequently resolve *Blastocystis'* potential impact in gastrointestinal health. A common method for elucidating the presence of these metabolites is through ^{1}H nuclear magnetic resonance (NMR). However, there are currently no described benchmarked methods available to extract metabolites from *Blastocystis* for ^{1}H NMR analysis. Herein, several extraction solvents, lysis methods and incubation temperatures were compared for their usefulness as an extraction protocol for this protozoan. Following extraction, the samples were freeze-dried, re-solubilized and analysed with ^{1}H NMR. The results demonstrate that carrying out the procedure at room temperature using methanol as an extraction solvent and bead bashing as a lysis technique provides a consistent, reproducible and efficient method to extract metabolites from *Blastocystis* for NMR.

Keywords: *Blastocystis*; ^{1}H NMR; metabolite extraction, metabolomics

Citation: Newton, J.M.; Betts, E.L.; Yiangou, L.; Ortega Roldan, J.; Tsaousis, A.D.; Thompson, G.S. Establishing a Metabolite Extraction Method to Study the Metabolome of *Blastocystis* Using NMR. *Molecules* **2021**, *26*, 3285. https://doi.org/10.3390/molecules26113285

Academic Editor: Robert Brinson

Received: 30 April 2021
Accepted: 26 May 2021
Published: 29 May 2021

Publisher's Note: MDPI stays neutral with regard to jurisdictional claims in published maps and institutional affiliations.

1. Introduction

Blastocystis is a genus of anaerobic protozoan that resides in the gastrointestinal tract of many vertebrate species and has historically been classified as a parasite, yet its pathogenicity has been a subject of dispute in recent years. *Blastocystis* has a unique metabolism and possesses a mitochondrial-related organelle (MRO) with chimeric characteristics of an aerobic mitochondrion and hydrogenosomes [1]. Many of these characteristics have been acquired by lateral gene transfer from prokaryotes and possibly other eukaryotic organisms in the gastrointestinal tract, and these have likely supported the adaptation of *Blastocystis* to the gut environment [2].

Previous in vitro studies aimed at mapping the unique metabolic pathways in *Blastocystis* have been based on genome and transcriptome analyses [3–5]. Biochemical analysis has involved fractionation, the separation of organelles by isopycnic density and the analysis of absorbance following the addition of certain substrates [6]. The latter of these approaches monitors enzyme activity in different organelles based on available nutrients and added substrates in vitro. This approach is limited in the range of enzymes and pathways that can be monitored. Therefore, a technique in which the whole metabolome can be analysed in the context of the host or in vitro culture is required. Metabolomics is a technique which can be utilised to analyse the metabolome of a cell or microorganism. This technique has been used to analyse the metabolomes of many microbes [7,8], plants [9],

83

nematodes [10] and animal cells [11–13]. Additionally, it has also been used to detect the molecules present in biological liquids such as blood [14], urine [14–16] and breast milk [17]. Mass spectrometry (MS) is probably the most popular analysis method for the detection and characterisation of small molecules and has been extremely successful because of its high sensitivity [10,18]. However, its arduous sample preparation can involve many steps to produce samples with good ionisation and MS properties. Subsequently, this can result in a loss of sample and the integrity of the metabolites being prejudiced. Therefore, reproducibility and accurate quantification can be difficult to achieve. In contrast, NMR can provide a simpler, more reproducible method for quantitative molecule detection, albeit with considerably lower sensitivity. NMR does not require the same laborious sample preparation that MS does, and the sample can remain intact throughout the analysis, thus making it a better quantitative tool [19–24]. However, for reasons of practicality and health and safety, NMR methods still require the extraction of metabolites from semi-solid samples such as cell cultures, as high resolution ^{1}H NMR is a solution state method. The question then becomes which solvent and method should be used to best isolate the desired group of molecules from a sample. For example, methanol is commonly used to extract polar molecules [10,11,13], while chloroform is commonly used to extract non-polar molecules [10,11].

Currently, the only protozoan parasite to have its metabolome analysed by NMR is *Giardia lamblia* [7]. In this study, the metabolome of *G. lamblia* was analysed by high resolution ^{1}H magic angle spinning (HR-MAS) NMR. HR-MAS does not require an extraction solvent as the cells remain intact [7]. However, HR-MAS experiments have some major drawbacks: firstly, they require a relaxation filter to exclude larger molecules such as proteins, as these produce a background unfavourable for the integration of sharper peaks, thus hampering quantification and comparison. [25]. The presence of this relaxation filter affects the sensitivity of the experiment and reduces the number of metabolites that can be detected. Secondly, HR-MAS experiments are limited by the volumes and quantities of samples that can be run with a maximum of 50 μL, which is at least ten times lower than the volumes usually used in liquid state NMR.

^{1}H NMR spectra have a proven track record for metabolite analysis from a number of biofluids and extraction methods [10–15,17]. Therefore, a combination of ^{1}H NMR metabolomics using a 1D-^{1}H-NOESY pulse sequence with an extraction protocol that only extracts small molecules provides an effective method for mapping *Blastocystis* metabolic pathways.

Herein, we aimed to investigate different extraction approaches in order to develop the optimum step-by-step method to extract metabolites from *Blastocystis* for analysis via ^{1}H NMR in order to analyse its metabolism.

2. Materials and Methods

2.1. Blastocystis Culture

Blastocystis ST7 cultures were grown axenically in 8 mL of Iscove's Modified Dulbecco's Medium (IMDM) (Gibco-Catalogue no 12,200,069 Thermo Fisher scientific) with 10% heat-inactivated horse serum (HIHS) (Gibco-Catalogue no 26,050,088 Thermo Fisher scientific). All cultures were passaged every 3–4 days depending on their growth rate and were subsequently expanded. All cultures were incubated at 37 °C in 95% CO_2 and 5% O_2. The gas concentration was maintained by a gas pack (BD-Catalogue no 261205) in an anaerobic chamber (Oxoid-Product code 10,107,992 Fisher scientific). Cell counts were achieved manually using a Neubauer haemocytometer (Brand-Catalogue no 717810).

2.2. Cell lysis and Metabolite Extraction

Blastocystis cultures intended for metabolite extraction were pooled in a 50 mL tube and centrifuged at 1000× g for 5 min at 4 °C and the supernatant was discarded. Resulting pellets were re-suspended in 5 mL of Locke's solution and given 2× washes with Stone's modification of Locke's solution (ATCC medium 1671), which was removed by a subsequent centrifugation at $1000 \times g$ for 5 min at 4 °C. The washed pellets were snap frozen in liquid nitrogen and stored at −80 °C.

Three steps were implemented for each experiment to determine the optimum extraction protocol and were each repeated four times. The conditions of each of the 4 experiments are shown in Table 1.

Table 1. Conditions of each experiment used to determine the best lysis method, incubation temperature and extraction solvent.

Experiment No.	Batch No.	Extraction Solvent	Lysis Method	Incubation Temp
1	1	4 mL EtOH (3:1) −20 °C	Sonication 3 × 30 s	3 min −20 °C
	2	4 mL MeOH (1:1) −20 °C		
2	1	4 mL MeOH (1:1) −20 °C	Bead Bashing–200 mg beads vortex 30 s	3 min −20 °C
	2	4 mL MeOH (1:1) −20 °C	Sonication 3 × 30 s	
3	1	4 mL MeOH (1:1) −20 °C	Sonication 3 × 30 s	3 min −20 °C
	2	4 mL MeOH (1:1) RT		3 min RT
4	1	4 mL MeOH (1:1) 60 °C	Sonication 3 × 30 s	3 min 60 °C
	2	4 mL MeOH (1:1) RT		3 min RT

Step 1: Three cell cultures were thawed, resuspended in 5 mL of Lockes' solution and then homogenised by vortexing for 30 s. These were then divided into two equal weight batches for parallel analysis. Each batch was centrifuged at $1000 \times g$ for 5 min at 4 °C, after which the supernatant was removed.

Step 2: The two batches were added to one of two different solvents: either 4 mL of ethanol:water (3:1) or 4 mL of methanol:water (1:1). The two different solvent batches were further processed at either −20 °C, room temperature (RT) or 60 °C (with samples for each solvent at each of the three temperatures). Each batch was then disrupted using one of two methods; either sonication in 3 × 30 s bursts or bead bashing by vortexing with 200 mg of 0.4 mm glass beads for 30 s followed by a 3-min incubation at either −20 °C, RT or 60 °C, then followed by vortexing for a further 30 s.

Step 3: Resulting solutions were then divided into 4 × 1 mL aliquots and centrifuged for 15 min at 4 °C at $10,000 \times g$. The supernatants were decanted into fresh tubes and lyophilised.

2.3. Preparation for ^{1}H NMR Acquisition

The lyophilised desiccates were suspended in 330 μL of milliQ H_2O, then vortexed for 30 s. The four supernatants of each sample where recombined and 147 μL of D_2O containing 5 mM of non-deuterated DSS was added, resulting in a final DSS concentration of 0.5 mM.

2.4. Analysis of Aqueous Extracts by ^{1}H NMR Spectroscopy

One-dimensional (1D) ^{1}H spectra were obtained using a 600 MHz Avance III NMR spectrometer (Bruker) with a QCI-P cryoprobe with experiments measured at an calibrated temperature of 298K. Temperatures were calibrated using the residual protonated peaks from MeOH in a D4-MeOH sample to avoid radiation damping effects from the high Q value of the QCI-P cryoprobe used [26,27]. For each sample, the spectrometer was locked to D_2O and the experiments were measured automatically using ICON NMR and a set of custom macros. Calibrations were carried out for each sample using a short excitation sculpting experiment; these included automated tuning and matching, measurement of the water offset and 90° pulse calibration, which was made using the stroboscopic nutation method of Wu and Otting [28]. The soft pulse power levels were calculated based on attenuated values calculated from the 90° pulse. The receiver gain measured for each sample and was limited to a maximum value of 128. A 1D-^{1}H NOESY 100 ms mixing time was run. Data were accumulated over 512 scans with eight dummy scans. A spectral width of 12.02 ppm (7211 Hz) was used, and 32,768 data points were acquired, giving an acquisition time of 2.27 s. Acquisitions were separated by a relaxation delay of 3 s. The relaxation delay was increased, and the acquisition time decreased to provide sufficient water suppression.

2.5. Processing and Analysis of ^{1}H NMR Data

All NMR spectra were phased, manually baseline corrected and exponentially line-broadened with a 1Hz window function using TOPSPIN 3.6.1 (Bruker) software. The spectra were then imported into Chenomx 8.4. A shim correction of 1.2 Hz was applied and the region from 4.56 ppm to 4.97 ppm was deleted to eliminate water resonance peaks. Peak assignment was performed using the Chenomx profiler tool fitting the spectral line to the proposed compounds in the standard Chenomx library. The efficacies of the extraction solvents, lysis methods and incubation temperatures were then compared using molecule concentration ratios and number of metabolites ratios between the two samples (Figure S3) (e.g.: $I_{E/M}$ = I ethanol/I methanol).

The median, standard deviation (StDev) and coefficient of variance (CV) were all calculated to determine the reproducibility of the results. Any outliers were detected and removed from the analysis.

3. Results

In order to determine the optimal protocol to extract metabolites from *Blastocystis* ST7 for NMR analysis, a series of extraction solvents, lysis techniques and incubation temperatures were examined. The efficacy of each protocol was assessed using proton NMR and the peak intensity was compared using TOPSPIN 3.6.1 to determine which method extracted the highest concentrations of metabolites. We then developed an efficient, reproducible protocol to perform metabolomics studies on *Blastocystis* species and found that the extraction solvent and lysis method were the most important factors for metabolite extraction. The efficacy was optimised in four sets of experiments, which firstly compared solvents (MeOH versus EtOH), then compared methods (sonication vs. bead bashing) and finally the temperature regime used ($-20\,°C$ versus RT,) and ($60\,°C$ versus RT).

3.1. Comparison of Steps

Two analysis methods were used during the comparisons of pairs of processing steps to rank efficacy. These were molar concentration ratios C $\mu M_{A/B}$ for processes A and B, as measured using the standard Chenomx metabolite library against the internal DSS standard. Secondly, the ratios of the raw number of detectable metabolites extracted $N_{A/B}$ using the two processes (A and B), again using the Chenomx metabolite library. All analyses were made in pairs of samples in triplicates, and samples in the triplicate were denoted by Arabic numerals 1–3 and condition pairs by A and B. Therefore, for a comparison of methanol and ethanol, 1A–3A were ethanol samples and 1B–3B were methanol samples.

3.2. Extraction Solvent

The first part of this investigation focused on determining the most suitable extraction solvent (ethanol or methanol) for the extraction of Blastocystis from cultures.

Two sets of triplicates of metabolite extractions from Blastocystis cells were trialled using ethanol or methanol as an extraction co-solvent with water. The efficacies of the extraction solvents were compared using C $\mu M_{A/B}$ and $N_{A/B}$ between the two samples calculating the ratio of ethanol/methanol. The ethanol extractions were labelled sample 1A–1C and methanol extractions were labelled sample 2A–2C. The results of the extractions are shown in Figure 1a,b as C $\mu M_{E/M}$ for a selected set of molecules and $N_{E/M}$, respectively. The triplicates shown in Figure 1a show that extraction from ethanol and water vs. extraction from methanol and water produced two consistent results. Four molecules from the 1A vs. 2A sample set were identified as outliers (Figure 1a). The 1A vs. 2A sample set was also identified as an outlier for the number of molecules extracted. The reproducibility of the triplicates was measured by the CV (Table S1, Supplementary Information) and the CV improved as the outliers were removed (Figure S1, Supplementary Information). All the reproducible results were below one, with the exception of formate and acetate in the sample set 1A vs. 2A and sample set 1A vs. 2A for the number of molecules extracted. The CV for the number of molecules extracted was 0.7, showing poor reproducibility. These results suggest that methanol worked better than ethanol. All six of the selected metabolites produced values below one in two of the three sample sets, and two of the three sample sets produced values below one for the number of metabolites extracted. Taken together, the results suggest that methanol was the better extraction solvent.

3.3. Lysis Method

The lysis method for metabolite extraction was subsequently investigated as part of this experiment; here, samples which had been extracted with methanol (deemed the most suitable extraction solvent) were subjected to different lysis techniques.

Two sets of triplicates of metabolite extractions from *Blastocystis* cells were examined with either bead bashing or sonication as the differing lysis methods. The efficacies of the lysis methods were compared using C $\mu M_{A/B}$ and $N_{A/B}$ between the two samples calculating the ratio of sonication/bead bashing. The sonicated extractions were labelled sample 3A–3C and bead-bashed extractions were labelled sample 4A–4C. The results of the extractions are shown in Figure 2a,b as 'C $\mu M_{S/B}$' for a selected set of molecules and 'N $_{S/B}$', respectively.

The triplicates show that for lysis by bead bashing vs. lysis by sonication, bead bashing produced more consistent results for the number of metabolites extracted, with all three triplicates being below one (Figure 2b). For the metabolite concentrations extracted, two metabolites were noted as outliers: alanine and formate for the pair sample set 3C vs. 4C (Figure 2a) and were removed and the CVs dropped from 0.7 to 0.02 and 0.6 to 0.03, respectively (Figure S2 and Table S2, Supplementary Information). All other peaks yielded three reproducible triplicates (Figure 2a). Of the reproducible triplicates, seven gave C $\mu M_{A/B}$ ratios which were below 1.0 and five that were above; these produced no significant results on aggregate. The number of molecules extracted produced reproducible triplicates (Figure 2b) with no outliers and a CV of 0.27 (Figure 2b), all of which were below a ratio of 1.0. These results suggest that bead bashing was a more reliable method for lysis of *Blastocystis* when compared to sonication.

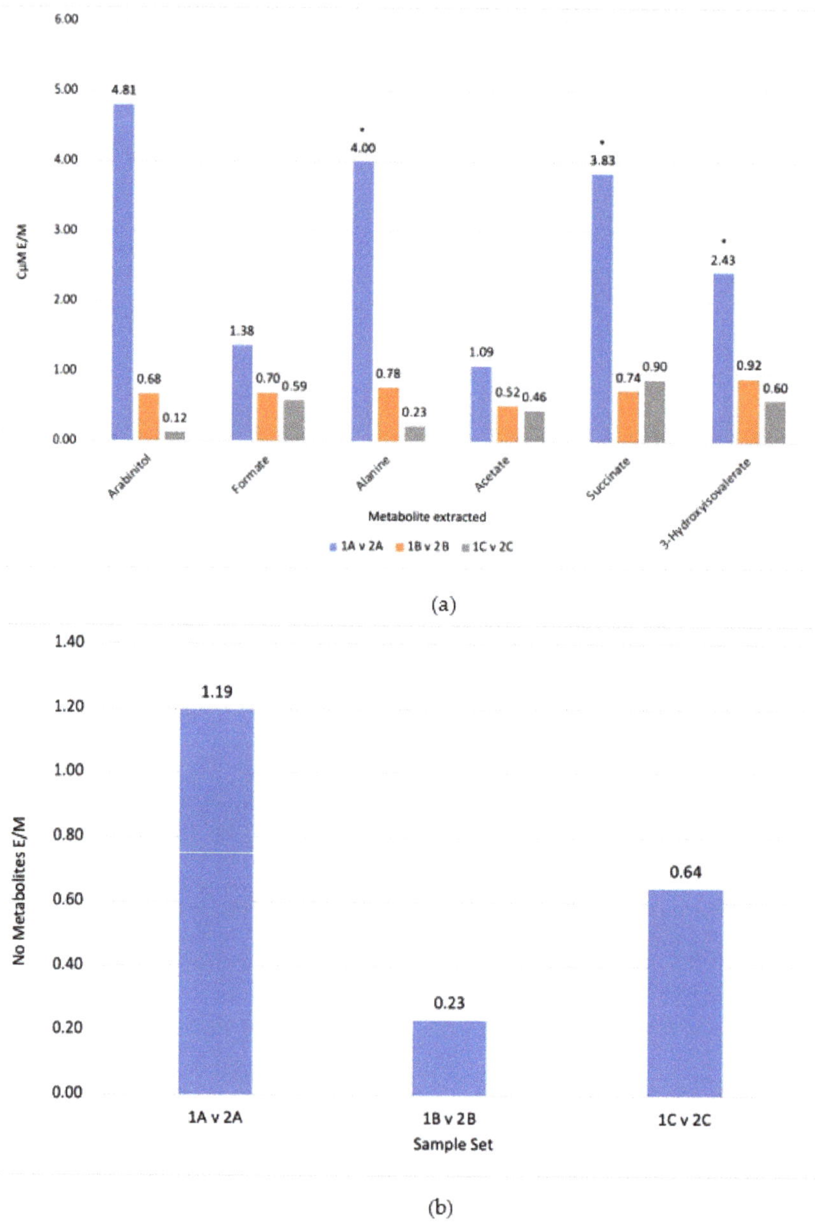

(a)

(b)

Figure 1. (**a**) Difference in metabolite concentrations between ethanol (1) and methanol (2) C μM $_{E/M}$ extractions for the triplicates A–C. (**b**) Difference in the number of different metabolites extracted between ethanol (1) and methanol (2) extractions $N_{E/M}$ for the triplicates. Numbers below 1.0 indicate an increased extraction in methanol, * = outliers, numbers above the bars indicate measured ratios.

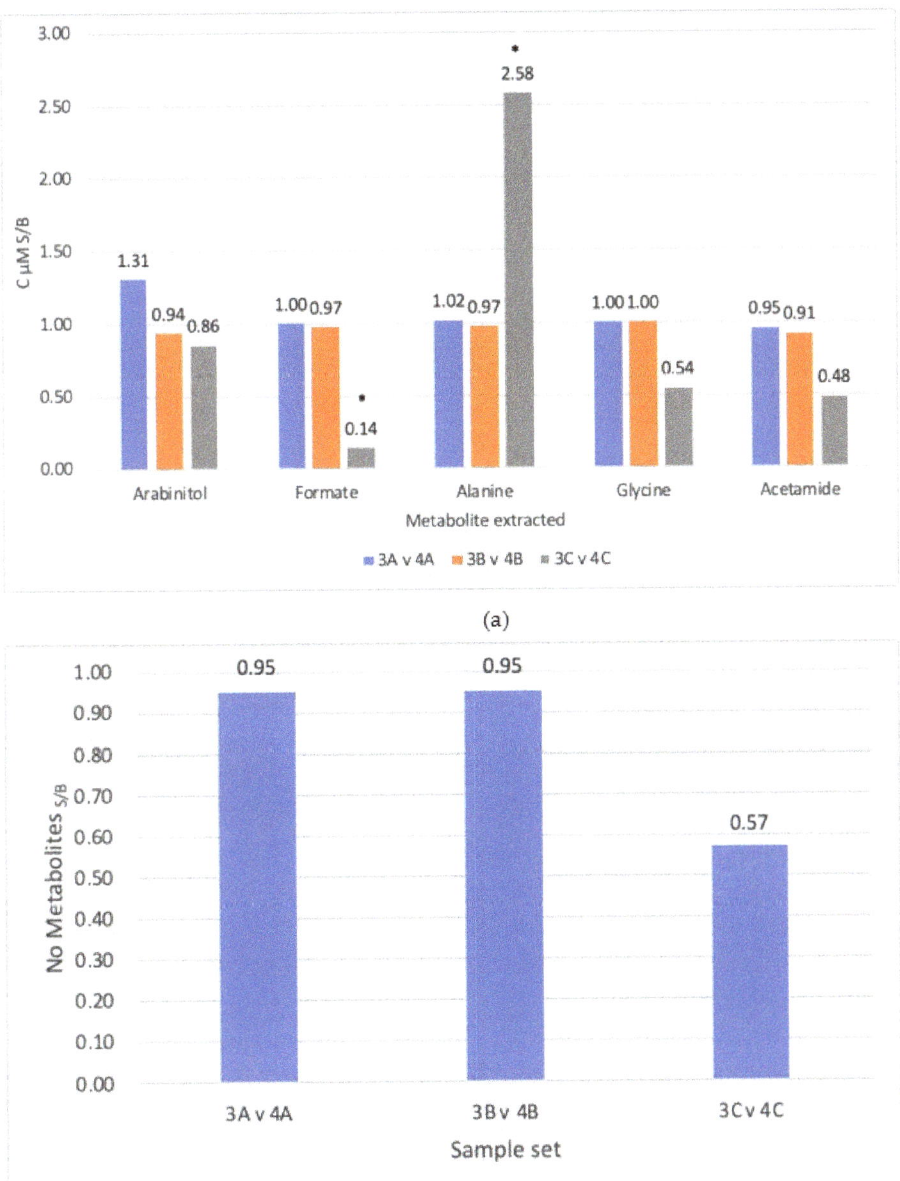

Figure 2. (**a**) Difference in concentrations between sonication (3) and bead bashing (4) C μMS/B lysis techniques for the triplicates A–C. (**b**) Difference in the number of different metabolites extracted between sonication (3) and bead bashing (4) lysis techniques NS/B for triplicates. Numbers below 1 indicate an increased extraction for bead bashing. * = outliers, numbers above the bars indicate measured ratios.

3.4. Incubation Temperature

Lastly, the final part of this investigation aimed at assessing the best incubation temperature for the extraction of metabolites from Blastocystis cultures. This part of the experiment used samples that had undergone extraction with methanol (extraction solvent) and bead bashing (lysis technique), chosen because they proved the most suitable methods, as described above.

Two sets of triplicates of metabolite extractions from *Blastocystis* cells were trialled under the following incubation temperatures: $-20\ °C$ or room temperature (RT). The efficacies of the incubation temperatures were compared using C $\mu M_{A/B}$ and $N_{A/B}$ between the two samples calculating the ratio of RT$/-20\ °C$. Results of the extractions are summarised in Figure 3a,c as C $\mu M_{RT}/-20\ °C$ for a selected set of molecules and $N_{RT}/-20\ °C$, respectively. The triplicates produced show that incubation at $-20\ °C$ vs. incubation at RT produced consistent results, with no outliers (Table S3, Supplementary Information). All but one of the result medians were within 0.1 of 1 (Table S3, Supplementary Information), meaning no significant results were produced. The number of molecules extracted also produced consistent results with a CV of 0.21, but there were no significant differences between the two temperatures. Therefore, neither temperature appeared to be the more efficacious for metabolite extraction.

In addition to investigating the effect of RT and $-20\ °C$ incubation temperatures, a $60\ °C$ incubation was also trialled. Two sets of duplicates of metabolites extracted from Blastocystis cells were included, using $-20\ °C$ or $60\ °C$ as the incubation temperatures. The efficacies of the incubation temperatures were compared using C $\mu M_{A/B}$ and $N_{A/B}$ between the two samples calculating the ratio of $60\ °C/-20\ °C$. The results of the extractions are shown in Figure 3b,d as C $\mu M\ 60\ °C/-20\ °C$ for a selected set of metabolites and $N_{60/-20\ °C}$, respectively. Duplicates were executed for this test and produced consistent results. The CVs all ranged between 0.01 and 0.29, suggesting that all results were reproducible (Table S4, Supplementary Information). There were no significant differences between the different extraction temperatures. Additionally, the number of metabolites extracted produced reproducible results, with a CV of 0.14 (Figure 3d).

Overall, it was determined that temperature was not an important factor in metabolite extraction here. This means that performing the experiment at RT would be sufficient to extract metabolites from Blastocystis.

The best extraction protocol (methanno/bead-bashing/RT) gave the 1D-^{1}H-NMR spectrum shown in Figure 4, with Table S5, Supplementary Information containing the list of the most abundant molecules identified in this spectrum. Arabinitol and formate were the most abundant molecules. However, amino acids such as alanine and leucine were also be identified, along with molecules involved in Blastocystis energy metabolism such as acetate and succinate. Small sugars such as disaccharide trehalose and monosaccharide galactitol were identified, along with the lipid membrane component sn-Glycero-3-phosphocholine. Other molecules with biological roles such as betaine and malonate were also detected. Betaine has a role in regulating osmotic stress.

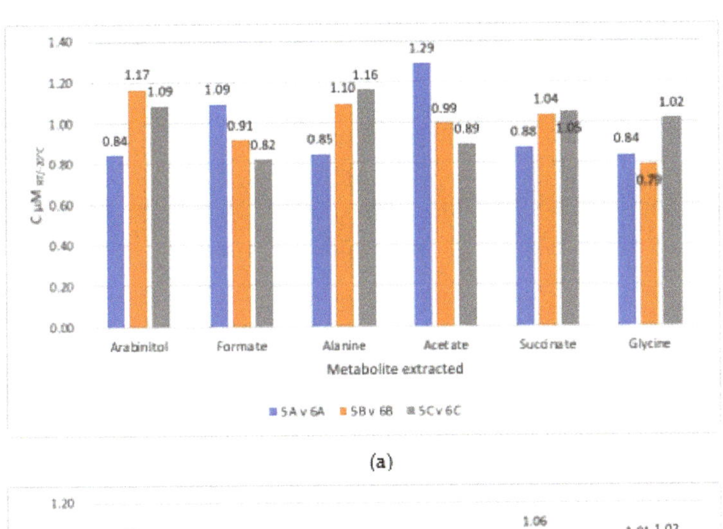
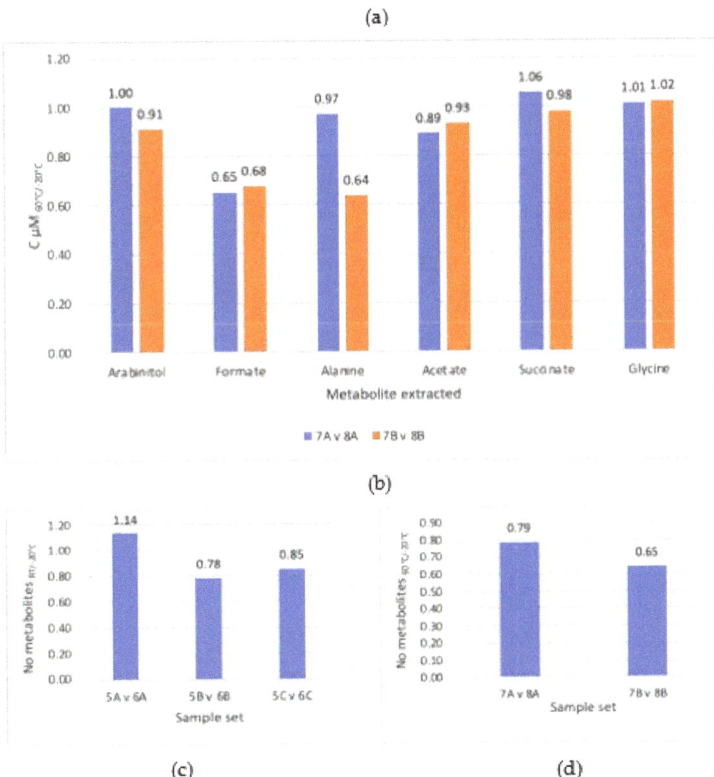

Figure 3. (**a**) Difference in concentrations between RT (5) and −20 °C (6) C µMRT/−20°C incubation temperature for triplicates A–C. Numbers below 1 indicate an increased extraction for −20 °C incubation. (**b**) Difference in concentrations between 60 °C (7) and −20 °C (8) incubation temperatures for triplicates A–C. Number below 1 indicate an increased extraction for −20 °C incubation. (**c**) Difference in the number of different metabolites extracted between RT (5) and −20 °C (6) incubation temperatures. NRT/−20 °C for triplicates A–C. Numbers below 1 indicate an increased extraction for −20 °C incubation (**d**) Difference in the number of different metabolites extracted between 60 °C (7) and −20 °C (8) incubation temperatures N60 °C /−20 °C for triplicates A–C. Numbers below 1 indicate an increased extraction for −20 °C incubation. Numbers above the bars indicate measured ratios.

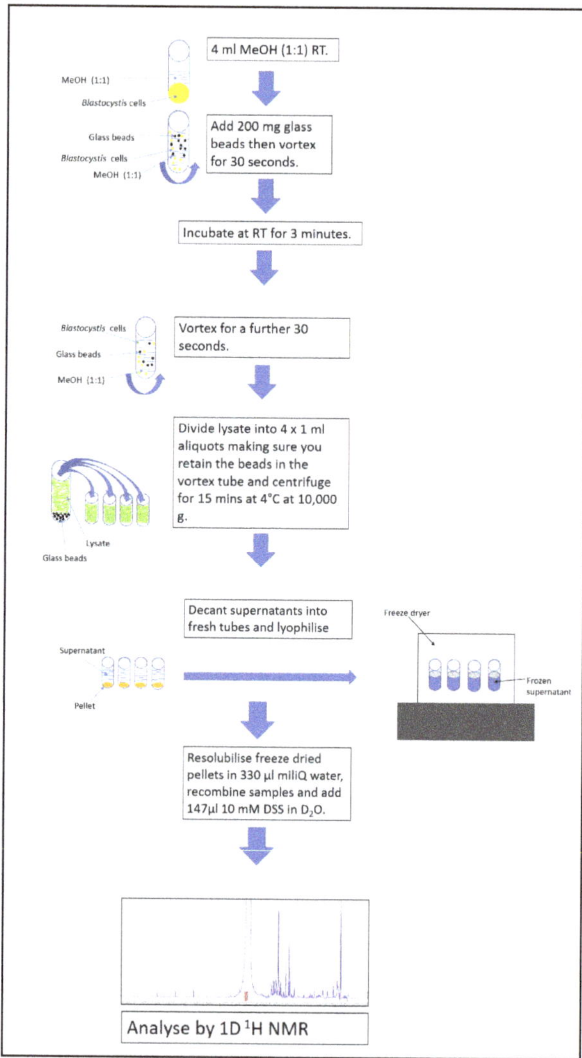

Figure 4. Final metabolite extraction protocol optimised by this study. Methanol is used as the extraction solvent, bead bashing as the lysis technique and incubation at RT.

4. Discussion

Herein, we have described an efficient protocol to extract metabolites from *Blastocystis* ST7 in culture, thus allowing an overview of its metabolome by ^{1}H-NMR analysis to be established for the first time. The findings can be summarized as follows: (1) methanol is a more effective extraction solvent when compared against ethanol; (2) bead bashing is a more effective lysis method than sonication; (3) incubation temperature is not a significant factor in metabolite extraction of *Blastocystis*; thus, performing the extraction at room temperature (RT) is sufficient. These data were collated to produce a series of steps to form an effective protocol to perform metabolite extraction on *Blastocystis* (Figure 4).

4.1. Methanol Was Determined to Be the Optimal Extraction Solvent

The results demonstrated that methanol was a more suitable solvent when compared against ethanol (Figure 1a). The molecule analysis produced six reproducible results: four of the molecules had one outlier and the other two had three reproducible results with one, which suggested ethanol was a better extraction solvent. All of the outliers and results that suggested ethanol was better came from a single sample (sample A). This could have been caused by an error in aliquot division when mixing a culture of cells, or homogeneity of the sample may not have been successfully achieved. The number of molecules extracted were consistent with metabolite concentration analysis, with sample set A being the only triplicate in which ethanol demonstrated better metabolite extraction than methanol. Overall, these results indicate that for *Blastocystis* ST7, methanol is a better extraction solvent. This is in contradiction to one past publication, in which a comparison between methanol and ethanol both produced similar results [29].

4.2. Bead Bashing Was Determined to Be the Optimal Lysis Method

Bead bashing was determined to be a more effective lysis technique when compared against sonication (Figure 2a). These are non-aggressive lysis techniques employed for *Blastocystis* as it does not possess a cell wall and is a single-celled organism, so cells are not connected by an extracellular matrix. One study by Geier et al. on *Caenorhabditis elegans* investigated different bead beating techniques, including some at cryogenic temperatures which produced successful results. A tissue homogenizer proved to be the most effective method here, yet it should be considered that *C. elegans* is a multicellular organism, meaning a more aggressive lysis technique is required [10]. Other research has demonstrated that cryopulveristation and tissue homogenisers were successful techniques for the lysis of mammalian cells [12,13]. However, sonication had proved successful in *Arabidopsis thaliana* [9], which has a cell wall and is tougher to break than *Blastocystis*. As sonication and bead bashing had both proved successful in tougher cells than *Blastocystis*, these two methods were selected. Bead bashing produced reproducible results (Figure 2a) against sonication, with only two selected peaks determined as outliers amongst all the samples. Nevertheless, the results of the extracted metabolite concentration ratios were not significant. The differences in concentrations of metabolite extracted ranged between 0.48 and 1.31 (Figure 2a) for most of the selected extracted metabolites, with the exception of formate and alanine in the 3C vs. 4C sample set, whose differences in concentration ranged between 0.14 and 2.58. The number of metabolites extracted produced three reproducible triplicates all suggesting that bead bashing was a better lysis technique than sonication and thus, bead bashing was consistently more successful than sonication.

4.3. Temperature Was Not an Important Factor in Metabolite Extraction

Incubation temperature was determined to not be a significant factor in successful metabolite extraction from *Blastocystis*. Additionally, as higher temperatures are more likely to facilitate chemical reactions, performing the experiment at room temperature may be essential for maintaining metabolite integrity. This is consistent with a past study by Beltran et al. [11]. However, it could also be the case that a 3-min incubation at the relevant temperature may not be long enough to have a sufficient effect and provides an avenue for future research into method optimisation. We would also like to emphasize that due to the nature and sensitivity of the organisms to oxygen, the objective was to minimise the extraction time to maintain sample integrity. RT against $-20\,°C$ (Figure 3a) produced a range of metabolite concentration ratios between 0.79 and 1.29. There were therefore no consistent, significant results and this was reproducible, suggesting that neither RT nor $-20\,°C$ was more successful. In past studies on human vein tissue and *C. elegans*, incubation at RT has been successfully performed [10,30], and similar experiments using *A. thaliana* demonstrated that successful extractions had been performed at $-20\,°C$.

In the $60\,°C$ incubation against the $-20\,°C$ incubation (Figure 3b), all of the extracted metabolite concentration ratios were between 0.64 and 1.06. All of the ratios were reproducible between the samples and there was no significant difference determined between them. For the number metabolites extracted ratios were both below 1.0, suggesting that −20 °C incubation was a more efficient incubation temperature to perform metabolite extraction than 60 °C. As RT was shown to be of similar efficacy to −20 °C, RT was selected as the best and most practical incubation temperature.

In summary, the most effective protocol determined by this study is shown in Figure 4. To summarize, this included methanol as the extraction solvent, accompanied by bead bashing and incubation at room temperature. Lyophilisation was used in each trial as a drying method and appeared to be a clean, consistent and successful drying technique. Although many of the results were reproducible, there were numerous outliers and, in some cases, only two reproducible results were produced amongst triplicates. For this reason, future work will aim to include more repeats in order to increase the reliability of the data. Therefore, for our final protocol quintuplets will be used, thus allowing the dismissal of one outlier, if necessary, to have successful triplicates.

The metabolites extracted by this protocol include amino acids such as alanine and leucine and molecules involved in energy metabolism such as acetate and succinate (Table S5, Supplementary Information). Additionally, a wide range of other molecules involved in biological processes such as betaine and malonate were present. The protocols trialled produced a range of metabolites numbering between 25 and 65. These were all polar molecules, as the solvents used target polar metabolites specifically. In the only other metabolomic study of a protozoan parasite, Vermathen et al. detected 31 different metabolites in *Giardia lamblia* using ^{1}H HR-MAS NMR. However, they detected 22 amino acids (18 proteogenic and 4 non-proteogenic) which is at a higher abundance than what was detected here in *Blastocystis* [4]. However, molecules such as betaine and succinate which are involved in biological processes were not detected in *G. Lamblia* [4] but were detected in quite a high abundance in *Blastocystis*. This could be because of the two organisms' different metabolisms, but also may be due to *Blastocystis* morphing into the cyst form and altering its metabolism subject to environmental changes.

Other NMR metabolomics studies of eukaryotic cells have demonstrated a similar number of metabolites to that extracted from *Blastocystis* at high concentrations. In a study on *Caenorhabditis Elegans* by Geier et al., 32 metabolites were detected at concentrations ranging between 2.48 mM and 5.73 mM [7]. Furthermore, in a study by Geier et al. on the avian liver, 52 polar metabolites were detected [10], and in a study on the rat liver by Lee et al., 30 metabolites were detected at concentrations ranging between 13.6 µM and 5.28 mM using methanol as an extraction solvent [8]. Bruno et al. extracted 38 metabolites from skeletal muscle using methanol and chloroform [9]. Methanol and chloroform form a two-layered solution with chloroform on top and methanol on the bottom. The polar metabolites migrate towards the methanol layer and the non-polar metabolites migrate towards the chloroform layer [9].

Even though we were unable to analyse a wider range of molecules, our established methodology was determined to be the most efficacious process from this study to use for the extraction steps for future metabolomics studies on *Blastocystis*. There are a wide range of metabolites which were not detected in this study which have been detected in past studies to map *Blastocystis'* metabolism. Malate, oxaloacetate and succinyl-coA, for example, are all involved in *Blastocystis* energy metabolism and ATP generation, but were not detected using this extraction method [1]. Additionally, the production of amino acids isoleucine and serine have also been detected in past studies [1] but were not detected using this method. This could be down to *Blastocystis* morphing into cyst form and its metabolism becoming dormant but could also be down to the inefficiency of this method for extracting those specific metabolites.

5. Conclusions

In this study, we developed an efficient and robust protocol to extract and analyse polar metabolites from *Blastocystis*. We generated many ^{1}H-NMR spectra to provide detail on the efficacy of each step of the protocol. This is the first extraction method described for NMR metabolomics analysis of *Blastocystis* species and it will spearhead future investigations to determine the metabolome of other *Blastocystis* subtypes, both in vitro, but also in vivo (e.g., stool metabolomic profiles). As such, this easy-to-use procedure could be applied to establish biomarkers in stool samples that could be subsequently used for (infectious) disease diagnosis.

Supplementary Materials: The following are available online. Table S1: C $\mu M_{E/M}$ for a selection of metabolites extracted from the ethanol vs. methanol experiment; Table S2: C $\mu M_{E/M}$ for a selection of metabolites extracted from the sonication vs bead bashing experiment; Table S3. C $\mu M_{E/M}$ for a selection of metabolites extracted from the RT vs. $-20\,^{\circ}$C incubation experiment; Table S4. C $\mu M_{E/M}$ for a selection of metabolites extracted from the $60\,^{\circ}$C vs. $-20\,^{\circ}$C incubation experiment; Table S5. Metabolites extracted by the optimal extraction method and their concentrations. Figure S1: Reproducibility of the ethanol extractions against the methanol extractions-C.Vs of ethanol vs. methanol metabolites extracted with outliers and without outliers; Figure S2: Reproducibility of sonication lysis against bead-bashing lysis. C.Vs of sonication vs. bead bashing metabolites extracted with outliers and without outliers. Figure S3. Spectrum obtained from the optimal extraction protocol deduced from this study.

Author Contributions: Conceptualisation, J.M.N. and A.D.T.; methodology, J.M.N., E.L.B., L.Y. and G.S.T.; software, G.S.T.; validation, G.S.T., E.L.B. and J.O.R.; formal analysis, J.M.N., L.Y. and E.L.B.; investigation, J.M.N.; resources, A.D.T.; data curation, J.M.N.; writing—original draft preparation, J.M.N.; writing—review and editing, G.S.T., J.O.R. and A.D.T.; supervision, G.S.T. and A.D.T.; project administration, A.D.T.; funding acquisition, A.D.T. All authors have read and agreed to the published version of the manuscript.

Funding: This research was funded by Biotechnology and Biological Sciences Research Council, grant number: BB/M009971/1. The NMR facility is supported by a Wellcome Trust Equipment Grant 091163/Z/10/Z (UK). J.M.N. was supported by a Kent-Health studentship and E.L.B. is supported by a GTA studentship from the School of Biosciences, University of Kent.

Institutional Review Board Statement: Not applicable.

Informed Consent Statement: Not applicable.

Conflicts of Interest: The funders had no role in the design of the study; in the collection, analyses, or interpretation of data; in the writing of the manuscript, or in the decision to publish the results.

Sample Availability: Not Available.

References

1. Stechmann, A.; Hamblin, K.; Pérez-Brocal, V.; Gaston, D.; Richmond, G.S.S.; van der Giezen, M.; Clark, C.G.; Roger, A.J. Organelles in *Blastocystis* That Blur the Distinction between Mitochondria and Hydrogenosomes. *Curr. Biol.* **2008**, *18*, 580–585. [CrossRef]
2. Tan, K.S.W. New Insights on Classification, Identification, and Clinical Relevance of *Blastocystis* spp. *Clin. Microbiol. Rev.* **2008**, *21*, 639–665. [CrossRef]
3. Denoeud, F.; Roussel, M.; Noel, B.; Wawrzyniak, I.; Da Silva, C.; Diogon, M.; Viscogliosi, E.; Brochier-Armanet, C.; Couloux, A.; Poulain, J.; et al. Genome Sequence of the Stramenopile *Blastocystis*, a Human Anaerobic Parasite. *Genome Biol.* **2011**, *12*, R29. [CrossRef]
4. Pérez-Brocal, V.; Clark, C.G. Analysis of Two Genomes from the Mitochondrion-Like Organelle of the Intestinal Parasite *Blastocystis*: Complete Sequences, Gene Content, and Genome Organization. *Mol. Biol. Evol.* **2008**, *25*, 2475–2482. [CrossRef]
5. Wawrzyniak, I.; Roussel, M.; Diogon, M.; Couloux, A.; Texier, C.; Tan, K.S.W.; Vivarès, C.P.; Delbac, F.; Wincker, P.; El Alaoui, H. Complete Circular DNA in the Mitochondria-like Organelles of *Blastocystis hominis*. *Int. J. Parasitol.* **2008**, *38*, 1377–1382. [CrossRef]
6. Lantsman, Y.; Tan, K.S.W.; Morada, M.; Yarlett, N. Biochemical Characterization of Amitochondrial-like Organelle from *Blastocystis* Sp. Subtype 7. *Microbiology* **2008**, *154*, 2757–2766. [CrossRef] [PubMed]
7. Vermathen, M.; Müller, J.; Furrer, J.; Müller, N.; Vermathen, P. 1H HR-MAS NMR Spectroscopy to Study the Metabolome of the Protozoan Parasite *Giardia lamblia*. *Talanta* **2018**, *188*, 429–441. [CrossRef]
8. Li, W.; Lee, R.E.B.; Lee, R.E.; Li, J. Methods for Acquisition and Assignment of Multidimensional High-Resolution Magic Angle Spinning NMR of Whole Cell Bacteria. *Anal. Chem.* **2005**, *77*, 5785–5792. [CrossRef] [PubMed]

9. Salem, M.A.; Jüppner, J.; Bajdzienko, K.; Giavalisco, P. Protocol: A Fast, Comprehensive and Reproducible One-Step Extraction Method for the Rapid Preparation of Polar and Semi-Polar Metabolites, Lipids, Proteins, Starch and Cell Wall Polymers from a Single Sample. *Plant Methods* **2016**, *12*, 1–15. [CrossRef]
10. Geier, F.M.; Want, E.J.; Leroi, A.M.; Bundy, J.G. Cross-Platform Comparison of *Caenorhabditis elegans* Tissue Extraction Strategies for Comprehensive Metabolome Coverage. *Anal. Chem.* **2011**, *83*, 3730–3736. [CrossRef]
11. Beltran, A.; Suarez, M.; Rodríguez, M.A.; Vinaixa, M.; Samino, S.; Arola, L.; Correig, X.; Yanes, O. Assessment of Compatibility between Extraction Methods for NMR- and LC/MS-Based Metabolomics. *Anal. Chem.* **2012**, *84*, 5838–5844. [CrossRef]
12. Bruno, C.; Patin, F.; Bocca, C.; Nadal-Desbarats, L.; Bonnier, F.; Reynier, P.; Emond, P.; Vourc'h, P.; Joseph-Delafont, K.; Corcia, P.; et al. The Combination of Four Analytical Methods to Explore Skeletal Muscle Metabolomics: Better Coverage of Metabolic Pathways or a Marketing Argument? *J. Pharm. Biomed. Anal.* **2018**, *148*, 273–279. [CrossRef]
13. Fathi, F.; Brun, A.; Rott, K.H.; Cobra, P.F.; Tonelli, M.; Eghbalnia, H.R.; Caviedes-Vidal, E.; Karasov, W.H.; Markley, J.L. NMR-Based Identification of Metabolites in Polar and Non-Polar Extracts of Avian Liver. *Metabolites* **2017**, *7*, 61. [CrossRef]
14. Chiu, C.Y.; Cheng, M.L.; Chiang, M.H.; Wang, C.J.; Tsai, M.H.; Lin, G. Metabolomic Analysis Reveals Distinct Profiles in the Plasma and Urine Associated with IgE Reactions in Childhood Asthma. *J. Clin. Med.* **2020**, *9*, 887. [CrossRef] [PubMed]
15. Leenders, J.; Grootveld, M.; Percival, B.; Gibson, M.; Casanova, F.; Wilson, P.B. Benchtop Low-Frequency 60 MHz NMR Analysis of Urine: A Comparative Metabolomics Investigation. *Metabolites* **2020**, *10*, 155. [CrossRef]
16. Saude, E.; Sykes, B. Urine Stability for Metabolomic Studies: Effects of Preparation and Storage. *Metabolomics* **2007**, *3*, 19–27. [CrossRef]
17. Gómez-Gallego, C.; Morales, J.M.; Monleón, D.; du Toit, E.; Kumar, H.; Linderborg, K.M.; Zhang, Y.; Yang, B.; Isolauri, E.; Salminen, S.; et al. Human Breast Milk NMR Metabolomic Profile across Specific Geographical Locations and Its Association with the Milk Microbiota. *Nutrients* **2018**, *10*, 1355. [CrossRef]
18. Zheng, J.; Mandal, R.; Wishart, D.S. A Sensitive, High-Throughput LC-MS/MS Method for Measuring Catecholamines in Low Volume Serum. *Anal. Chim. Acta* **2018**, *1037*, 159–167. [CrossRef] [PubMed]
19. Emwas, A.-H.; Roy, R.; McKay, R.T.; Tenori, L.; Saccenti, E.; Gowda, G.A.N.; Raftery, D.; Alahmari, F.; Jaremko, L.; Jaremko, M.; et al. NMR Spectroscopy for Metabolomics Research. *Metabolites* **2019**, *9*, 123. [CrossRef] [PubMed]
20. Giraudeau, P. NMR-Based Metabolomics and Fluxomics: Developments and Future Prospects. *Analyst* **2020**, *145*, 2457–2472. [CrossRef]
21. Markley, J.L.; Brüschweiler, R.; Edison, A.S.; Eghbalnia, H.R.; Powers, R.; Raftery, D.; Wishart, D.S. The Future of NMR-Based Metabolomics. *Curr. Opin. Biotechnol.* **2017**, *43*, 34–40. [CrossRef]
22. Nagana Gowda, G.A.; Raftery, D. Can NMR Solve Some Significant Challenges in Metabolomics? *J. Magn. Reson.* **2015**, *260*, 144–160. [CrossRef]
23. Takis, P.G.; Ghini, V.; Tenori, L.; Turano, P.; Luchinat, C. Uniqueness of the NMR Approach to Metabolomics. *TrAC Trends Anal. Chem.* **2019**, *120*, 115300. [CrossRef]
24. Vignoli, A.; Ghini, V.; Meoni, G.; Licari, C.; Takis, P.G.; Tenori, L.; Turano, P.; Luchinat, C. High-Throughput Metabolomics by 1D NMR. *Angew. Chem. Int. Ed.* **2019**, *58*, 968–994. [CrossRef]
25. Palmioli, A.; Ceresa, C.; Tripodi, F.; La Ferla, B.; Nicolini, G.; Airoldi, C. On-Cell Saturation Transfer Difference NMR Study of Bombesin Binding to GRP Receptor. *Bioorganic Chem.* **2020**, *99*, 103861. [CrossRef] [PubMed]
26. Findeisen, M.; Brand, T.; Berger, S. A 1H-NMR Thermometer Suitable for Cryoprobes. *Magn. Reson. Chem.* **2007**, *45*, 175–178. [CrossRef] [PubMed]
27. Raiford, D.S.; Fisk, C.L.; Becker, E.D. Calibration of Methanol and Ethylene Glycol Nuclear Magnetic Resonance Thermometers. *Anal. Chem.* **1979**, *51*, 2050–2051. [CrossRef]
28. Wu, P.S.C.; Otting, G. Rapid Pulse Length Determination in High-Resolution NMR. *J. Magn. Reson.* **2005**, *176*, 115–119. [CrossRef]
29. Snytnikova, O.A.; Khlichkina, A.A.; Sagdeev, R.Z.; Tsentalovich, Y.P. Evaluation of Sample Preparation Protocols for Quantitative NMR-Based Metabolomics. *Metabolomics* **2019**, *15*. [CrossRef]
30. Anwar, M.A.; Vorkas, P.A.; Li, J.V.; Shalhoub, J.; Want, E.J.; Davies, A.H.; Holmes, E. Optimization of Metabolite Extraction of Human Vein Tissue for Ultra Performance Liquid Chromatography-Mass Spectrometry and Nuclear Magnetic Resonance-Based Untargeted Metabolic Profiling. *Analyst* **2015**, *140*, 7586–7597. [CrossRef] [PubMed]

 *molecules*

Article

Personalized Metabolic Profile by Synergic Use of NMR and HRMS

Greta Petrella [1], Camilla Montesano [2], Sara Lentini [1], Giorgia Ciufolini [1], Domitilla Vanni [1], Roberto Speziale [3], Andrea Salonia [4,5], Francesco Montorsi [4,5], Vincenzo Summa [3], Riccardo Vago [4,5], Laura Orsatti [3], Edith Monteagudo [3] and Daniel Oscar Cicero [1,*]

[1] Department of Chemical Science and Technology, University of Rome "Tor Vergata", 00133 Rome, Italy; petrella@scienze.uniroma2.it (G.P.); sara.lentini84@gmail.com (S.L.); ciufolini@scienze.uniroma2.it (G.C.); vanni@scienze.uniroma2.it (D.V.)

[2] Chemistry Department, University of Rome "Sapienza", 00185 Rome, Italy; camilla.montesano@uniroma1.it

[3] IRBM S.p.A., 00071 Pomezia, Italy; r.speziale@irbm.com (R.S.); vincenzo.summa@unina.it (V.S.); l.orsatti@irbm.com (L.O.); edith.monteagudo@chdifoundation.org (E.M.)

[4] Urological Research Institute, IRCCS Ospedale San Raffaele, 20132 Milan, Italy; salonia.andrea@unisr.it (A.S.); montorsi.francesco@unisr.it (F.M.); vago.riccardo@hsr.it (R.V.)

[5] Division of Experimental Oncology, URI Urological Research Institute, IRCCS San Raffaele Scientific Institute, 20132 Milan, Italy

* Correspondence: cicero@scienze.uniroma2.it

Abstract: A new strategy that takes advantage of the synergism between NMR and UHPLC–HRMS yields accurate concentrations of a high number of compounds in biofluids to delineate a personalized metabolic profile (SYNHMET). Metabolite identification and quantification by this method result in a higher accuracy compared to the use of the two techniques separately, even in urine, one of the most challenging biofluids to characterize due to its complexity and variability. We quantified a total of 165 metabolites in the urine of healthy subjects, patients with chronic cystitis, and patients with bladder cancer, with a minimum number of missing values. This result was achieved without the use of analytical standards and calibration curves. A patient's personalized profile can be mapped out from the final dataset's concentrations by comparing them with known normal ranges. This detailed picture has potential applications in clinical practice to monitor a patient's health status and disease progression.

Keywords: nuclear magnetic resonance; mass spectrometry; urine metabolome; normal ranges; personalized metabolic profile

Citation: Petrella, G.; Montesano, C.; Lentini, S.; Ciufolini, G.; Vanni, D.; Speziale, R.; Salonia, A.; Montorsi, F.; Summa, V.; Vago, R.; et al. Personalized Metabolic Profile by Synergic Use of NMR and HRMS. *Molecules* **2021**, *26*, 4167. https://doi.org/10.3390/molecules26144167

Academic Editor: Robert Brinson

Received: 14 June 2021
Accepted: 7 July 2021
Published: 8 July 2021

Publisher's Note: MDPI stays neutral with regard to jurisdictional claims in published maps and institutional affiliations.

1. Introduction

Over the last hundred years, biochemical discoveries have made it increasingly possible to characterize the metabolic pathways in our bodies, develop new drugs, and monitor human nutrition and lifestyle. Though our knowledge is increasingly broad, it remains divided into specific areas, such as the characterization of genetic make-up or transcriptional factors underlying the expression of essential proteins involved in specific physiological or pathophysiological processes. In this context, metabolomics has come into its own to mend the cracks between the different disciplines hitherto used to study our biochemical mechanisms [1]. Considering that a single change in a DNA base can lead to the observation of alterations in metabolite concentrations of up to 10,000-fold changes [2], metabolomics represents a highly sensitive probe for depicting our phenotype. Helped by the development of new analytical technologies for obtaining and processing biochemistry data, metabolomics as an omics discipline is under constant development. In the last 20 years alone, more than 5000 papers have been published on the subject, making it one of the fastest-growing disciplines [3].

The most used analytical platforms in metabolomics are chromatography-mass spectrometry (LC–MS, GC–MS, CE–MS, and IMS–MS) and NMR spectroscopy. As reported in many papers, these two methodologies have several features that make them complementary [4]. For example, MS techniques are highly sensitive and allow for the detection of thousands of features at different concentration ranges, potentially expanding the description of a metabolic profile in detail with just a few microliters of sample. However, the identification of compounds by MS is a more complex process than by NMR. Indeed, the metabolite identity is solved by measuring the mass-to-charge ratio (m/z) of the ionized molecule and/or its ionized molecular fragments and then comparing them with reference spectra and/or using analytical standards [5]. Furthermore, not all MS techniques have the same degree of reproducibility; this is mainly the case for LC–MS measurements, which yield less reliable metabolite quantifications [6].

Unlike mass spectrometry, NMR is not a destructive technique and, in many cases, requires minimal sample preparation. The ability to determine the identity of a compound with a single analysis (^{1}H-NMR) can be very accurate and fast for concentrated compounds or those that give signals in non-crowded regions of the spectrum. However, it has a lower sensitivity than MS [3], making it possible to only quantify a portion of the metabolome.

Thus far, few studies have described a combined use of both techniques, and these have directed their attention towards the development of statistical methods for weighing the two datasets [4] or for the structure determination of new compounds in commonly studied biofluids [7]. However, given the high complementarity of the two techniques, it should be beneficial to combine the data separately obtained with NMR and MS to improve the ability to classify and quantify the "metabotypes" under investigation [8].

One of the main limitations of using NMR is the relatively low number of accurately quantifiable metabolites in particularly complex mixtures like urine. For example, a study performed at two different fields (600 and 700 MHz) starting from a list of 151 metabolites that are potentially quantifiable in urine showed that only 50 presented data strongly correlated between the values obtained at the two magnetic fields [9]. This result represents a limit of quantifiable compounds in urine using NMR, and most studies in the literature have used a dataset of this size [10–15]. However, in one case, it was possible to reach 209 quantified metabolites [16], but only a fraction was detected in more than 80% of the samples.

Simultaneously identifying a metabolite by both NMR and MS would maximize the advantages for biomarker discovery by increasing the number of quantified metabolites in all samples and the accuracy of the measured concentrations. In our opinion, the method that has best combined MS data with NMR is the one developed by Nicholson et al. [17]. The authors named this strategy Statistical Heterospectroscopy (SHY) and showed that it is possible to correlate chemical shift and m/z data when a cohort of samples is considered. This concept revealed a new perspective to cross-reference NMR and MS data and to get the best of both techniques. However, the correlation was attempted with regions of the NMR spectrum, limiting the number of identifiable metabolites and obtaining only relative levels instead of concentrations. Our idea is to use the SHY concept to develop a novel strategy of the MS-assisted deconvolution of NMR spectra to extend the number of urinary metabolites quantified in their absolute rather than relative levels. We show how the synergistic use of both analytical methodologies can help to achieve this goal, taking the determination of metabolite concentrations in human urine as a specific case. We call this approach: *SY*nergic use of NMR and HRMS for *MET*abolomics (SYNHMET). Using SYNHMET, it was possible to obtain a complete dataset comprising 165 urinary metabolite concentrations for nine controls, six patients affected by chronic cystitis, and thirty-one bladder cancer patients.

2. Results

2.1. Metabolite Levels in Urine Acquired by NMR and HRMS

The SYNHMET method was first applied to quantify metabolites in urine samples from 46 subjects, divided into three groups: nine healthy controls (CTRL), six patients with chronic cystitis (CC), and thirty-one bladder cancer patients (BC). The ^{1}H-NMR dataset was acquired using a 600 MHz Bruker Avance spectrometer (Bruker, MA, USA). The HRMS dataset was acquired using a UHPLC–high-resolution mass spectrometry (UHPLC–HRMS) analysis system coupled to an Orbitrap QExactive™ mass spectrometer (Thermo Scientific™, MA, USA) equipped with a HESI source operated in the positive and negative ion modes. In addition, we used two different chromatographic conditions: reverse-phase (RP), which allows for the separation of metabolites based on hydrophobic interactions, and hydrophilic interaction liquid chromatography (HILIC), enabling the analysis of polar compounds. Combining two ion modes and two chromatographic conditions allowed for broad coverage of urinary metabolites, an essential feature in an untargeted approach. After the MS analysis, 10,497 hits were obtained. Information about a matched formula, exact mass, retention time, and relative intensity was available for each hit, and in many cases, so was a putative name (Table 1).

Table 1. Examples of two hits obtained with UHPLC–HRMS and the information available for each of them.

Mode [1]	Name	Formula	MW	Rt [min]	S_1	S_2	S_{46}
HC+	Caffeine	C8 H10 N4 O2	194.0805	4.636	2.36	2.81	0.85
RP-	-	C11 H15 N2 P	206.0976	9.727	0.39	0.36	0.14

[1] HC+: HILIC positive; RP-: reverse phase negative.

2.2. Extraction of Urinary Metabolite Concentrations by SYNHMET

The proposed workflow for the SYHNMET method applied to the human urine samples described above is shown in Scheme 1. The following chapters illustrate its application focused on a specific region of the ^{1}H-NMR spectrum.

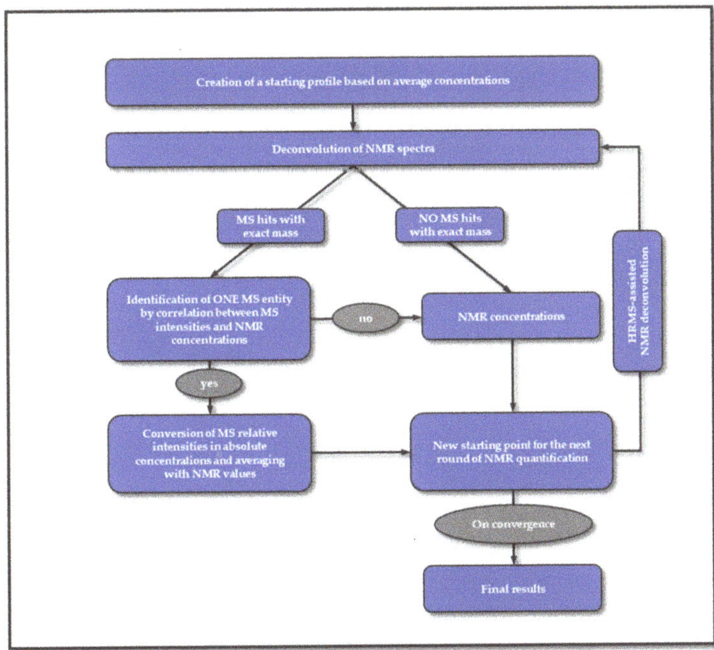

Scheme 1. Workflow of the SYNHMET method that allowed for the identification and accurate quantification of a significant number of metabolites in 46 urine samples in this application.

2.2.1. Creation of a Starting Profile and Deconvolution of NMR Spectra

Metabolites present in urine were first quantified by NMR using a deconvolution process. When this approach is applied, the goal is to minimize the difference between the experimental and calculated profiles. The latter is obtained by adding signals belonging to all the mixture components, weighted by their concentrations. To obtain a reasonable starting point for the first calculation, we selected 180 metabolites previously identified and quantified by NMR, considering their chemical shifts from the Chenomx database and their reported average concentrations in urine [16].

The SYNHMET procedure is illustrated using, as an example, the region between 2.47 and 2.37 ppm. The observed NMR profile of this zone mainly consists of the superimpositions of signals belonging to eleven metabolites: 2-oxoglutarate, 3-hydroxy-3-methylglutarate, 3-hydroxybutyrate, 4-pyridoxate, carnitine, glutamine, glutaric acid monomethyl ester, levulinate, pyroglutamate, succinate, and trans-4-hydroxy-L-proline.

In the first step, all chemical shifts and concentrations were changed to minimize the difference between calculated and experimental shapes. As a result, we obtained a list of metabolite concentrations constituted by very approximate values, especially those present at low levels or have their resonances hidden by other signals. The next step used MS-derived information to improve the accuracy of these measurements.

2.2.2. Using the NMR/MS Correlation to Identify an MS peak(s)

The first step to incorporate the HRMS measurements involved creating a list for each metabolite containing all MS-detected peaks showing a difference lower than 5 ppm between their measured accurate masses and the monoisotopic molecular weight. After this search, all the eleven metabolites were linked to a variable number of MS-chromatographic peaks, ranging from seventeen (glutamine and 4-pyridoxate) to one (3-hydroxybutyrate). These numbers show the level of ambiguity in identifying an MS hit based only on the exact mass. We exploited the correlation between the MS intensities and the NMR concentrations obtained in the first step to assist MS identification.

Such an MS feature selection process is illustrated in Figure 1 for 2-oxoglutarate, which shows the correlations between NMR concentrations obtained in the first deconvolution round and five different chromatographic peak intensities. Despite the expected inaccuracy of these NMR concentrations, we could still identify the HC- peak at 4.57 min as 2-oxoglutarate (Figure 1e).

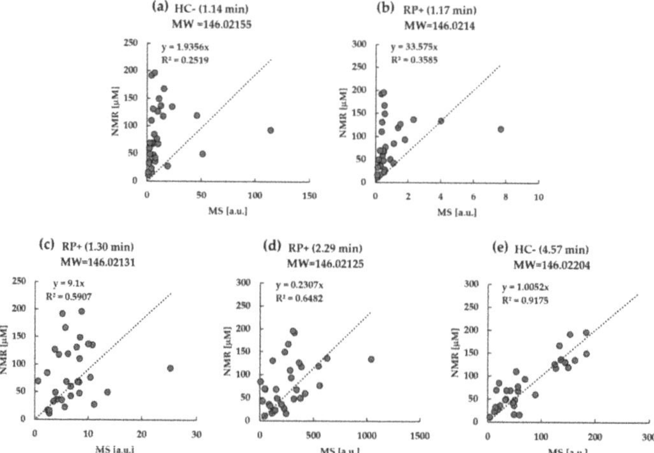

Figure 1. (a–e) Correlation plots between the relative MS intensities and 2-oxoglutarate NMR concentrations obtained after the first round for five peaks showing an exact mass compatible with the monoisotopic molecular weight of 2-oxoglutarate (monoisotopic MW = 146.021523302).

Using this procedure, we could unambiguously identify at least one MS feature for all other metabolites of this region, except for 3-hydroxybutyrate and trans-4-hydroxy-L-proline.

2.2.3. HRMS Assisted NMR Deconvolution

The MS intensities of the assigned chromatographic peaks were converted into concentrations using the slope of the linear correlations. These values were then averaged with those measured by NMR, employed to update the profiles obtained after the first round, and finally applied these profiles as starting points for the second round. Finally, the second round of deconvolution was performed.

Figure 2a,b follows the evolution of the calculated profile for two samples after the first (upper panel) and second-round (lower panel) of deconvolution. Main variations are evident for low concentrated metabolites, such as 4-pyridoxate (blue) and 2-oxoglutarate (purple).

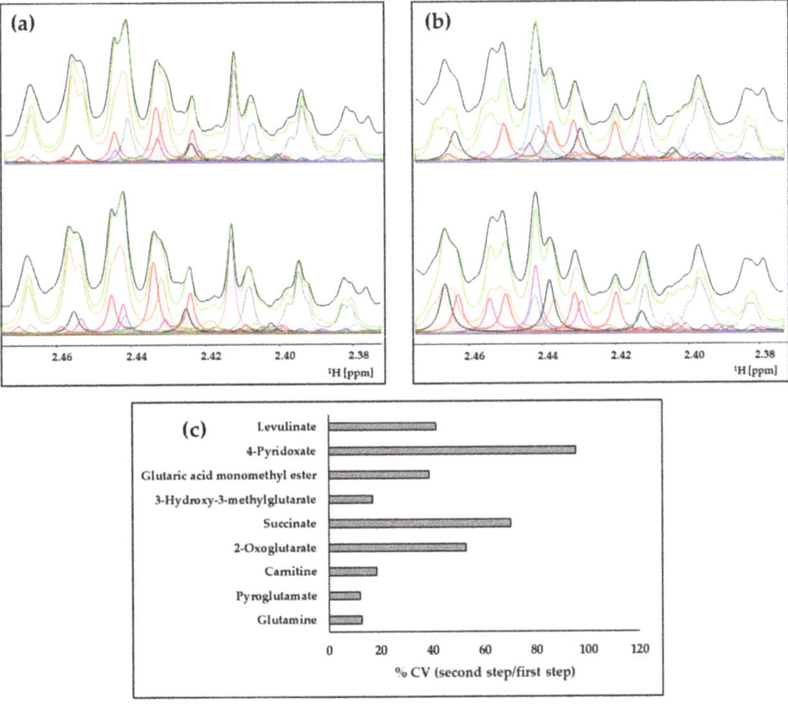

Figure 2. Example of the deconvolution process for two different samples (**a,b**) in the zone between 2.47 and 2.37 ppm. The first and second deconvolution steps are shown in the upper and lower panels for each spectrum, respectively. Black and green lines represent the experimental and calculated spectra, respectively. Signals from glutamine (orange), 2-oxoglutarate (light purple), carnitine (red), 3-hydroxy-3-methylglutarate (black), 4-pyridoxate (blue), trans-4-hydroxyproline (light green), pyroglutamate (gray), glutaric acid monomethyl ester (dark green), succinate (pink), levulinate (dark purple), and 3-hydroxybutyrate (electric blue) are shown. (**c**) Percent coefficients of variation (%CV).

One thing worth noting is that there was no significant improvement in the agreement between the calculated (green) and experimental (black) profiles after the second round compared to the first one, although the metabolite concentrations on which they were based were very different. Because there were multiple ways to replicate the profile by combining the levels and positions of the metabolite signals, a precise reproduction of the measured spectrum did not warrant obtaining accurate concentrations. A deconvolution process assisted by the HRMS values would likely yield the most reliable results.

The ratios between first- and second-round concentrations were first calculated for each of the 46 samples. Next, the percent coefficients of variation (%CV) for each metabolite were calculated to assess the degree of change in concentration of this set of metabolites after mass-assisted deconvolution (Figure 2c). Metabolites showing low changes are the only ones that NMR can reliably quantify. On the other hand, the high percentage of variation for many metabolites shows the extent to which it is necessary to cross-reference the data between NMR and MS to obtain accurate results relying on two orthogonal measurements.

2.2.4. Final Results

Table 2 summarizes the results obtained after two deconvolution rounds for the eleven metabolites clustered into the region of 2.47–2.37 ppm. Nine out of eleven metabolites could be unambiguously linked to one or more MS hits. Due to the different chromatographic conditions and polarities used, a single compound could be represented by more than one peak, such as for 4-pyridoxate, carnitine, and succinate. The final concentrations were calculated for all these nine metabolites by averaging the values obtained from the MS and NMR measurements, which showed a significantly improved correlation after the second round. This increase in the R^2 value was a natural consequence of using the deconvolution process assisted by UHPLC–HRMS intensities, but its final value was still a measurement of the degree of agreement between these two datasets measured orthogonally. For 3-hydroxybutyrate and trans-4-hydroxy-L-proline, we were not able to identify any MS hits correlating with the NMR concentrations. In the first case, the final concentrations were still measured by NMR because we judged their values to be sufficiently reliable. On the contrary, the concentrations of trans-4-hydroxy-L-proline were not included in our final results. This compound has signals with high multiplicity, which divides the intensity into multiple components and does not present any resonance in a non-crowded region of the spectrum. These two facts make the NMR measurements inaccurate, hindering our ability to identify the correct MS hit(s) among the eleven showing its exact mass.

This procedure was repeated for all the regions into which the NMR spectrum was divided. As a result, we were able to quantify 165 metabolites out of the 180 initially considered. Of this total, twelve were quantified using only NMR data. Our final concentration matrix contained only 48 missing values, representing 0.6% of the total. These concentrations were normalized by converting them into μM/mM of creatinine, and they were compared with those of the literature (Table S1). The excellent agreement between the retention times of nine labeled standards co-injected with the samples with those obtained by the HRMS-NMR correlation method (Table S2) further sustains the assignments of the metabolites listed in Table S1.

The set of metabolites quantified cover a wide range of biochemical markers, including amino acids and their metabolism, markers of vitamins, dysbiosis, diet and toxin exposure, carbohydrates and their metabolism, energy, fatty acid/lipid, and glycine/serine metabolism, and ketone bodies. In this way, it is possible to cover some of the central metabolic pathways for metabolomics studies to discover biomarkers related to pathological states and individual profiling.

Table 2. Identified and quantified metabolites belonging to the example region between 2.47 and 2.37 ppm. The table shows for each compound the diagnostic chemical shifts for their identification and quantification, the number of chromatographic peaks detected in the mass dataset, the chromatographic peak(s) identified by the correlation with NMR data, and the value of Pearson's correlation coefficient in the first and second deconvolution step.

Metabolite	δ [ppm]	Chr. Peaks [1]	Chr. Peaks [2] (RT [min])	R^2 Initial [3]	R^2 Final [4]
2-Oxoglutarate	2.4, 3.0	5	HC- (4.57)	0.90	0.97
3-Hydroxy-3-methylglutarate	1.3, 2.4	15	HC- (3.69)	0.96	0.99
3-hydroxybutyrate	1.2, 2.4	1	-	-	-
4-Pyridoxate	2.4, 7.9	17	RP+/− (4.41), HC− (1.09)	0.86, 0.87, 0.89	0.98, 0.98, 0.97
Carnitine	3.2, 2.4	9	RP+(0.88), HC+ (8.02)	0.94, 0.95	0.97, 0.99
Glutamine	2.1, 2.4	17	RP+ (0.81)	0.97	0.99
Glutaric acid monomethyl ester	2.4	10	RP− (5.28)	0.90	0.98
Levulinate	2.2, 2.4	3	RP+ (1.11)	0.93	0.99
Pyroglutamate	2.4	12	RP+ (2.48)	0.96	0.99
Succinate	2.4	6	RP+/− (2.91), HC− (1.95)	0.99, 0.99, 0.98	0.99, 0.99, 0.99
trans-4-Hydroxy-L-proline	2.1, 2.4, 4.3	11	-	-	-

[1] Number of chromatographic peaks found in the MS dataset. Due to the different chromatographic conditions and polarities used, a single compound could be represented by more than one peak. [2] Chromatographic peak/s identified by correlation. Pearson correlation after [3] first round and [4] second round.

2.3. The Origins of the Analytical Synergism between NMR and UHPLC–HRMS

Many metabolites are difficult to quantify by NMR, mainly those in low concentrations and those whose signals are hidden by the presence of other resonances. We have defined this interference as the "NMR matrix effect" because its consequence is the same as observed when ESI is used as the ionization source [18]. Its impact differs from one spectrum to another, mainly in biofluids like urine that show highly variable composition. The use of MS intensities characterized by an exact mass and measured after a chromatographical separation highly alleviates this difficulty because the quantification is performed on separated components of the mixture. Combining the NMR and HRMS datasets offers the opportunity to obtain accurate concentrations for many metabolites, a result that would be impossible to achieve when using NMR alone.

However, if, on the one hand, the use of UHPLC–HRMS data expands the number of quantified metabolites, on the other hand, NMR aids in increasing the accuracy of MS-derived concentrations. The most frequent causes of error in the evaluation of concentrations by MS are the detector's saturation and the matrix effect. These two effects are unpractical to correct when a large matrix composed of a considerable number of samples and metabolites is analyzed in an untargeted MS-based analysis.

An example of the first case was found during the quantification of hippuric acid, which shows a wide range of concentrations in urine [19,20]. Figure 3 shows that the response of the MS detector was not linear for concentrations above 3.8 mM. Thus, we only considered NMR values for those samples with values above this limit to avoid significant errors.

Specifically, for the BC group, 82 values were found to lie outside the literature ranges. Most abnormal values corresponded to dietary components, followed by metabolites belonging to fatty acids/lipids, carbohydrates, energy, and branched-chain amino acid metabolisms. Nine metabolites previously found significantly altered in BC patients— namely O-acetylcarnitine, gluconate, lactate, phenylacetylglutamine, citrate, hippurate, succinate, valine, and erythritol [26]—were also found outside their normal ranges (Figure 4). The complete metabolic profile of one BC patient is shown in Figure S1. Twenty-four metabolic concentrations lay outside the literature ranges (Figure 5). They primarily belonged to components of the diet, fatty acid metabolism, and energy metabolism. These results underline the degree of detail that can be achieved with the SYNHMET methodology, with a potential clinical practice application to monitor apatient's health status and disease progression.

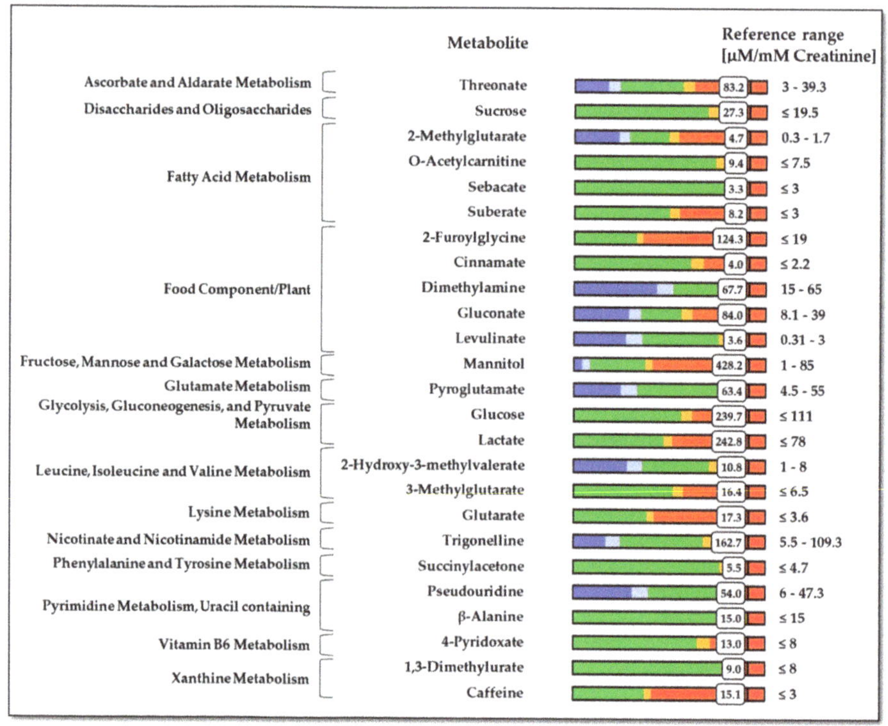

Figure 5. Urinary metabolites of a BC patient showing abnormal values according to literature ranges. Blue and red areas represent 10% lower and higher values than those reported in the literature for adults over 18 years old, respectively. All values are expressed in μM/mM of creatinine.

3. Discussion

The value of combining the two most commonly used techniques in metabolomics, NMR and MS, was recently recognized and addressed in a review by Marshall and Powers [8]. However, no method has attempted to directly correlate an NMR chemical shift with an MS *m/z* value of a single sample because there is no specific information to indicate that these two features belong to the same molecule [8]. MS intensities belonging to a cohort of samples were cross-correlated with NMR spectral regions to overcome this limitation [17]. The correlation that does not exist in one sample exists in all of them as a group because, at this point, it is the distribution of intensities that determines whether a given chemical shift belongs to a molecule signal presenting a certain m/z. The so-called Statistical Heterospectroscopy approach, however, led to the identification of a reduced number of metabolites. The major drawback of this approach is, in our opinion, that the correlation was attempted between intensities of compounds whose levels are measured separately by HPLC–MS with regions of the NMR spectrum whose intensities result from the simultaneous contributions of many metabolites. A clear correlation between the MS and the NMR bin intensities can only be expected for strongly dominating metabolites because of their concentration in the region's shape.

Differently, SYNHMET uses the resolution power of NMR to separate most of the different signals contributing to the spectrum profile, coupled to that of UHPLC–HRMS. The deconvolution strategy was used to extract more than 200 metabolite concentrations from urine [16], a result not reproduced in any further study to the best of our knowledge. The difficulty associated with this methodology lies mainly in the extraction of levels for not concentrated metabolites or those presenting signals in crowded regions. These areas only provide the sum of the contributions of the various compounds, and without further information, there are many ways to combine the positions and intensities of the mixture components to reproduce the experimental shape of the NMR spectrum. The simultaneous use of UHPLC–HRMS intensities provides the key to obtain a single solution because it adds two new features to calculate the relative contribution of metabolites to a profile: the molecular weight and the chromatographic resolution. The latter is not practical in NMR measurements due to a combination of low sensitivity and long acquisition times. In this way, the correct proportions are extracted by combining NMR and UHPLC–HRMS, which transforms the experiment used for metabolite identification/quantification from monodimensional (chemical shift) into three-dimensional (by adding retention time and exact mass). Globally, the mechanism by which this method operates can be defined as an MS-assisted NMR deconvolution, improving the quality and quantity of the obtained data compared to that expected when exclusively using NMR (Scheme 2).

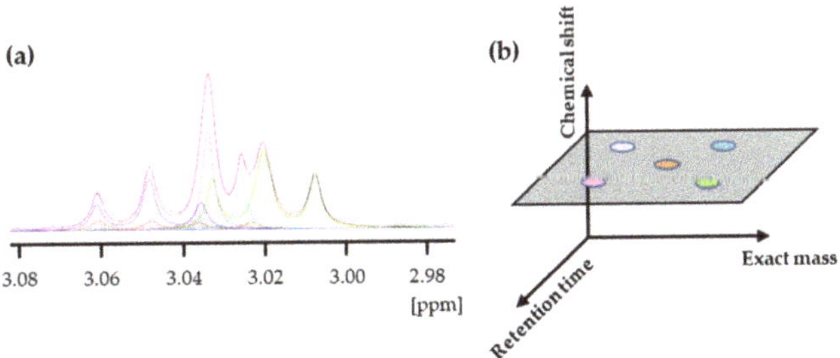

Scheme 2. Correlation between NMR concentration and MS intensity helps determine the relative position in crowded regions of the NMR spectrum, like in the example (**a**) that shows the superposition of five different signals corresponding to different metabolites by resolving the peaks in the retention time-exact mass plane (**b**).

According to the Metabolomics Standard Initiative, a definite metabolite identification, called level 1, needs a direct comparison of experimental data with an authentic reference standard [27]. It was argued that NMR metabolite identification of compounds in mixtures achieved by comparing with a spectra database approaches level 1 identification [27]. In the SYNHMET strategy, we added parameters characterizing a compound (chemical shift, multiplicity, and the number of signals) and the elemental composition provided by the correlation with the MS data to the NMR. This additional information limits the possible structures to the existing isomers, constituting a very restricted chemical space for low molecular weight compounds. The probability that two isomers show the same NMR parameters is extremely low, if even possible. For all these reasons, the confidence in the SYNHMET identification should be considered, in our opinion, similar to that in level 1.

Applying SYNHMET enabled us to quantify a large number of metabolites in urine. Many papers have supported the concept that the utility of a given approach is directly proportional to the measurable number of metabolite levels. However, this is only one of the two essential parameters in defining the value of a dataset for metabolomics studies. The other is the completeness of the matrix because if there are too many missing values, the classification ability or the detection of correlations between metabolites becomes weaker [28]. In our experience and from analyzing the literature on NMR urine metabolomics, the maximum number of metabolites quantified in at least 80% of samples is around 50–60 [10–15]. This number is far from that achieved in LC-MS studies, which have reached more than a thousand [29]. However, the identity of many of these metabolites is only putative because it is only supported by fragmentation spectra.

In our scheme, both the reproducibility and accuracy of the results are mainly supported by the characteristics of NMR. It is commonly accepted that these are two main robust features of NMR, which involve the possibility of obtaining the same instrumental response even when different spectrometers are used. These characteristics have favored constructing a community-built reference calibration line, with the participation of twenty-three laboratories, including ours [30]. We foresee that a similar calibration line can be produced among laboratories keen to prove the validity of the SYNHMET approach, allowing for the scientific community to obtain more robust results in metabolomics. However, the number of samples analyzed in this work did not allow for a definitive answer, and further studies will be needed to assess the limits in terms of precision and accuracy.

A compelling application of SYNHMET is the possibility of generating a detailed personalized profile of urinary metabolites. The main way to get a reliable profile is the election of an effective way to normalize the metabolites' concentrations to correct the variation induced by the subject hydration status. Usually, concentrations are normalized by the total urine volume collected during 24 h or the urinary creatinine level. These two normalization strategies present advantages and drawbacks. In case of using the total urine volume for 24 h, the incomplete collection is the main problem [22]. On the other hand, creatinine concentration is affected by several factors that are not directly related to the glomerular filtration rate, like muscle mass, diet, age, sex, and race [31,32]. Creatinine is also secreted from the renal tubules, which is not desirable for a glomerular filtration marker. A study comparing the uncertainties related to standardization of urine samples with volume and creatinine concentration showed that the latter introduces a 19–35% error [22]. However, if compared with the total volume normalization, this is partially counteracted by the higher risk that the sample is incomplete in collecting voids during a 24-h time interval. More recently, a study showed that normalization with urinary creatinine is better than volume in rats under controlled preclinical conditions, even when compared to a more recently proposed normalizer, cystatin C [33].

Our study used creatinine to normalize the metabolite concentrations, mainly because almost all the available normal ranges found in the literature are expressed in μM/mM of creatinine [34]. Independently on the used strategy, the profile of normalized metabolite concentrations constitutes a personalized urinary picture, which can be used to expand the current capability of classical biochemical tests to determine aperson's health status. This approach is very different from classical metabolomics, which seeks to find universal biomarkers of a disease or drug effects. The concept of personalized medicine grew up from the scientific evidence that there is high interindividual variability in the metabolic response to any change in the health status or the response to a drug. Therefore, expanding the number of metabolites that can be routinely monitored in biofluids can define a more accurate picture to be used in clinical practice [35]. At the heart of this analysis is the concept that a person's metabolic profile can reflect an individual's overall health status. Nowadays, physicians only capture a tiny fraction of the information contained in the metabolome, mainly due to its high complexity and the lack of robust and efficient analytical methods to determine the absolute instead of the relative level of a large number of chemical compounds in biofluids. Routine analyses only evaluate a very restricted number of compounds, such as glucose level for monitoring diabetes, cholesterol and low/high-density lipoproteins for cardiovascular health, or urea and creatinine for renal disorders. Simultaneously determining the absolute concentration of hundreds of molecules will open up new scenarios towards more accurate personalized medicine and increase the predictive value of such analyses.

For example, a patient suffering from BC showed a urinary profile with significant abnormal values for metabolites belonging to galactose/starch sucrose, caffeine, and lysine metabolisms (Figure 5). A recent study about recognizing different stages of BC using machine learning identified the first two as the main dysregulated metabolisms in early stages, whereas lysine metabolism was found to be unbalanced in late stages [36]. The case of caffeine metabolism is remarkable. Along with one of its metabolites, 1,3-dimethylurate, caffeine is processed by a P450 family cytochrome acting in the liver, CYP1A2 [37]. The connection between caffeine metabolism, exposure to tobacco compounds, and urinary mutagenicity has been known for a long time [38]. Significantly, cigarette smoking is the leading risk factor for BC, accounting for 50% of the total [39]. In addition to this patient, urine caffeine levels were significantly elevated in six other subjects with BC.

This patient also presented significant comorbidity due to cardiovascular pathologies, particularly severe myocardial ischemia. We observed different altered metabolisms related to cardiopathies, like those corresponding to branched-chain amino acids, lactate, and fatty acid metabolism [40]. They are the consequences of increased fatty acid metabolism, decreased glucose metabolism, and impaired branched-chain amino acid catabolism. Finally, the patient showed chronic pancreatitis, probably related to past alcohol abuse. The malfunction of the pancreas should explain the very high level of glucose in the urine, as in diabetic subjects.

4. Materials and Methods

4.1. Chemicals and Reagents

All used solvents and reagents were LC–MS grade. Water (H_2O), acetonitrile (ACN), formic acid (FA), and ammonium formate (CAS 540-69-2) were obtained from Sigma Aldrich (St. Louis, MO, USA). The stable isotope-labeled (SIL) internal standard $^{13}C^{15}N_2$-8-hydroxy-2′-deoxyguanosine ($^{13}C^{15}N_2$-8-OH-dG) was obtained from Toronto Research Chemicals (Toronto, ON, Canada). $^{15}N_4$-hypoxanthine ($^{15}N_4$-Hyp), L-tyrosine-(phenyl-d_4) (d_4-L-Tyr), and $^{15}N_4$-inosine ($^{15}N_4$-I) were purchased from Cambridge Isotope Laboratories, Inc., (Tewksbury, MA, USA). L-kynurenine sulfate: H_2O (ring-d_4, 3,3-d_2) (d_6-KYN) and D_2O were acquired from Cambridge Isotope Laboratories, Inc. (Andover, MA, USA). Anthranilic acid-ring-$^{13}C_6$ ($^{13}C_6$-AA) and 3-(trimethylsilyl)-2,2,3,3-d propionic acid (TSP) were purchased from Sigma Aldrich (Schnelldorf, Germany). $^{15}N,^{13}C_2$-3-Hydroxy-DL-kynurenine (^{15}N-$^{13}C_2$-OH-KYN) was obtained from AMRI (Albany, NY, USA).

4.2. Urine Collection

Urine samples were obtained from the Urological Research Institute (URI) of San Raffaele Hospital (Milan, Italy). Caucasian patients aged between 32 and 90 years were recruited. The dataset comprised 46 samples: 31 bladder cancer (BC) patients, nine healthy controls, and six with chronic cystitis. BC patients with concomitant or previous prostate, renal, or upper excretory tract cancer; urinary tract infections; or kidney failure were excluded. Urine samples were collected before the surgical intervention and processed soon after. The samples were centrifuged at 300 g for 5 min, aliquoted, and stored at –80 °C until use.

4.3. UHPLC–High Resolution Mass Spectrometry Analysis

4.3.1. SIL-Stock and Working Solution Preparation

Stock solutions were prepared from the independent weight of compounds and stored at −20 °C. d_6-KYN and $^{13}C_6$-AA were prepared in H_2O/DMSO (1/1, *v/v*) at 1.5 and 5.0 mg/mL, respectively. ^{15}N-$^{13}C_2$-OH-KYN was prepared in H_2O/DMSO (1/19, *v/v*) at

2.0 mg/mL. $^{13}C^{15}N_2$-8-OH-dG, $^{15}N_4$-Hyp, d_4-L-Tyr, and $^{15}N_4$-I were prepared in water at 1.0 mg/mL.

Internal Standard Working Solutions (IS-WS) were prepared by adding appropriate volumes of the stock solutions to 50 mL of ultrapure H_2O (ISWS-A) and ACN (ISWS-B) to reach a final concentration of 200 ng/mL for all the standards. The solutions were maintained at 4 °C and freshly prepared every week.

4.3.2. Urine Normalization by Specific Gravity

Specific gravity (SG) measurements were made with a portable digital refractometer (Atago UG-α, Tokyo, Japan). The refractometer had a urinary SG range from 1.000 to 1.060 with a resolution of 0.001. Urine samples were thawed at room temperature in an ultrasonic bath for 10 min and then centrifuged (4000 rpm). An aliquot of urine (100 μL) was placed upon the lens of the refractometer previously calibrated with LC–MS-grade water to measure SG values. Samples were then split into two aliquots. Urinary metabolite levels were normalized by SG-diluting each aliquot with water or ACN:H_2O in variable amounts for RP and HILIC analysis, respectively. Dilutions were performed to bring all samples to the same specific gravity value.

4.3.3. Urine Samples Preparation

All samples were further diluted by 3-fold with ISWS-A for RP analysis or ISWS-B for HILIC analysis. Samples were vortexed and centrifuged (13,000 g for 10 min), and the supernatant (350 μL) was transferred to a 96-well plate and randomized for LC–MS analyses.

4.3.4. Quality Control Samples and Blanks Preparation

Two different types of quality control (QC) samples were prepared: pooled QCs made by mixing equal volumes (5 μL) from each sample previously normalized for the specific gravity and dilution QCs prepared by 2, 4, and 8-fold diluting the pooled QCs with LC–MS-grade water. All QCs were further diluted by 3-fold with ISWS-A for RP analysis or ISWS-B for HILIC analysis. Pooled QC samples were injected first ($n = 20$) to condition the LC–MS system and obtain stable retention times and MS response. Subsequently, pooled QCs were injected every six true samples ($n = 8$ in total) to perform intra-batch signal drift corrections. Dilution QCs were analyzed four times and were regularly incorporated along the sample list to verify the linear response of the MS signal. Blanks consisted of LC–MS-grade water for RP analysis and ACN: H_2O 80:20 (v/v) for HILIC analysis. Blank injection ($n = 3$) was performed at the beginning of the batch to collect a background signal excluded from the dataset.

4.3.5. HILIC and RP Chromatography

The used UHPLC system was an Ultimate 3000™ liquid chromatographic system (Thermo Scientific™, MA, USA) coupled to an Orbitrap Q Exactive™ mass spectrometer (Thermo Scientific™, MA, USA) equipped with a HESI source operating in the positive and negative ion modes. HILIC chromatographic separation was accomplished using a BEH-HILIC column, 130 Å, 1.7 μm, and 2.1 × 100 mm (Waters, Milford, MA, USA). The used mobile phases were: 20 mM ammonium formate along with 0.1% FA at pH 3.7 (mobile phase A) and ACN (mobile phase B). The gradient consisted of a linear increase of mobile phase B from 5% to 35% over 8.5 min, followed by an additional increase to 50% in 1 min. Phase B was kept constant for 1.5 min and then decreased to 5% in 0.5 min and kept stable for 3.5 min for column re-equilibration (total run time of 15 min). The used flow rate was 0.300 mL/min, the injection volume was 2 μL, and the column was kept at 35 °C.

RP chromatographic separation was achieved using an HSS-T3 column, 100 Å, 1.7 μm, and 2.1 × 100 mm (Waters, Milford, MA, USA). The mobile phases were: 0.1% FA in H_2O (mobile phase A) and 0.1% FA in ACN (mobile phase B). The gradient ramp consisted of a linear increase to 10% of mobile phase B over 6 min and to 35% in 2 min. Mobile phase

B was further increased to 98% in 2 min, kept constant for 0.5 min, and finally decreased to 0% in 0.5 min and kept stable for 3 min for column re-equilibration (total run time of 15 min). The flow rate was 0.300 mL/min from 0 to 8.0 min, increased to 0.4 mL/min from 8.0 to 12.0 min for column washing, and brought back to 0.3 mL/min from 12.0 to 15.0 min. The injection volume was 2 µL, and the column was kept at 35 °C. During LC–MS analysis, samples were kept in the autosampler at 8 °C.

4.3.6. High-Resolution Mass Spectrometry

Mass spectra were acquired on an Orbitrap QExactive™ mass spectrometer (Thermo Scientific™, MA, USA) operating in both the positive and negative ion modes. The HESI parameters were: 3.20 kV (pos)/−3.20 kV (neg) electrospray voltage, 280 °C heated capillary temperature, 50 (pos)/−50 (neg) S-lens RF level, sheath gas (N_2) flow of 50 a.u., auxiliary gas (N_2) flow of 10 a.u., and gas temperature of 300 °C. The acquisition range was set from m/z 60 to 900 at a resolution of 70,000 FWHM at m/z 200. All data were acquired in profile mode using Xcalibur™ 3.1.66.10. The QExactive™ mass spectrometer was calibrated for the positive and negative modes before sample analysis using the calibration solution provided by the manufacturer (Pierce LTQ ESI Positive Calibration Solution and Pierce LTQ ESI Negative Calibration Solution). For the mass calibration of the instrument, a custom list that included lower masses than the default calibration provided with the instrument was used to ensure that accurate masses were detected at low molecular weights.

4.3.7. Raw Data Processing by Compound Discoverer

The raw files obtained in the positive and negative ion modes were separately processed using Compound Discoverer™ 2.0 (Thermo Scientific™). Four output tables (RP+, RP-, HILIC+, and HILIC-) were generated, including m/z, retention time, and peak intensity, for all the analyzed samples. An untargeted metabolomics workflow for retention time alignment, component detection, elemental composition prediction, and gap-filling was used. The workflow tree included the following nodes: input files, select spectra, align retention times, detect unknown compounds, group unknown compounds, fill gaps, normalization areas, and mark background compounds. The raw files were aligned with an adaptive curve setting with a 5 ppm mass tolerance and a 0.4 min retention time shift. Unknown compounds were detected with a 5 ppm mass tolerance, signal to noise ratio of 3, 30% of relative intensity tolerance for isotope search, and 10,000 minimum peak intensity, and then they were grouped with 5 ppm mass and 0.3 min retention time tolerances. A procedural blank sample was used for background subtraction and noise removal during the pre-processing step. Peaks were removed from the list if they showed less than a 3-fold increase compared to blank samples or if they were detected in less than 50% of QCs and/or with relative standard deviation (%RSD) of the QCs greater than 50%. To balance differences in intensities that may have arisen from instrument instability, a normalized area across all samples was provided for each detected metabolic feature by normalization to the periodically analyzed QC samples (pooled QC).

Finally, the hit intensities of each sample were multiplied by the dilution factor used for pre-normalization. Thus, un-normalized data were used to ensure a better degree of correlation between NMR and MS.

4.4. H-NMR Spectroscopy

4.4.1. Sample Preparation

The urine samples, previously stored at −80° C, were thawed on ice and centrifuged at 4000 rpm for 10 min at 4 °C; then, 500 µL of supernatant were collected. Then, 50 µL of phosphate buffer solution [41] (1.5 M K_2HPO_4/NaH_2PO_4, 30 mM NaN_3, and 5.5 mM TSP, pH 7.4 in D_2O) were added, and 50 µL of the final solution were transferred to a 1.7 mm thin-walled glass NMR tube for subsequent NMR analysis.

4.4.2. Spectra Acquisition

^{1}H-NMR experiments were performed on Bruker Avance 600 MHz equipped with a SampleJet autosampler using a noesypr1d sequence, mixing time of 100 ms, a spectral window of 12 ppm, acquisition time of 2 s, relaxing time of 3 s, 516 scans, 4 dummy scans, and T = 298 K. This sequence has become the best choice for NMR-based metabolomics studies [42] for several reasons. Firstly, the quality of water suppression is very high without the need for extensive optimization. Secondly, an increasing number of well-established groups utilize the sequence, reflecting its consistency [43]. Finally, the library of Chenomx used in this study to quantify metabolite concentrations is optimized for this sequence and compensates for incomplete relaxation.

4.4.3. H-NMR Data Analysis

All the spectra were processed using 0.5 Hz of line-broadening followed by manual phase and baseline correction. Chenomx NMRSuite 8.5 (Chenomx Inc.) was used to quantify the concentrations of the metabolites. The spectra database in this software allows for the manual deconvolution of different signals and determines the concentration of the compounds that form the mixture. TSP was set as an internal standard at 0.5 mM.

4.5. SYNHMET Method

The starting spectrum profile for deconvolution is defined using the average concentrations of urine metabolites [16]. The chemical shifts and levels of all compounds are then varied to reproduce the profile observed in each experimental NMR spectrum. The matching between the calculated and experimental spectral profiles is never perfect. The source of this inequality can be understood by analyzing all variables contributing to the spectrum intensity at a given chemical shift (I_k) (Equation (1)):

$$I_k \;=\; \sum_{i=1}^{n} a_i K_{i,k} \;+\; \sum_{j=1}^{m} b_j U_{j,k} \;+\; N_k \tag{1}$$

where k is the chemical shift, i represents one assigned metabolite, n is the total number of assigned metabolites, $K_{i,k}$ is a known factor accounting for the shape of assigned metabolites, j represents one unassigned metabolite, m is the total number of unassigned metabolites, $U_{j,k}$ is an unknown factor considering the shape of unidentified metabolites, a_i and b_j are the metabolite concentrations, and N_k is a random factor representing the noise.

In parallel, the exact mass of each metabolite is searched in the MS dataset, creating a list of linked MS features for most compounds. The number of MS peaks associated with each metabolite varies from zero to more than twenty. Detecting more than one peak with the same exact mass turns the identification based solely on the molecular weight uncertain unless using labeled standards. In the SYNHMET method, combining the concentrations measured for a cohort of samples simultaneously by MS and NMR can solve this ambiguity in an alternative way. We considered that a certain MS-detected chromatographic peak showing the accurate mass of a metabolite can be attributed to it when it is the only one showing a significant correlation between the distributions of the MS peak intensities and NMR concentrations. The intensities of the selected peak are then converted into concentrations by multiplying them with the slope of the best fit solution. The initial spectrum profile is then adjusted, inserting the values of the peak or peaks averaged to those measured by NMR for each metabolite. Conversely, concentrations of compounds not represented by any MS feature or showing multiple or no correlations are not updated for the following phase.

During the next profiling step, all compounds' signal positions and concentrations defining the updated profile are varied to obtain the best accordance between the calculated and experimental profiles. After completion, a new correlation test is accomplished, possibly increasing the number of identified and consequently quantified metabolites. This process is iteratively repeated until no further information is added. The final matrix contains concentrations of metabolites that are determined by a combination of MS and NMR measurements.

5. Conclusions

In conclusion, the new methodology for merging NMR and UHPLC–HRMS produced a list of 165 metabolite concentrations in urine in almost all samples, with significantly higher accuracy of identification and quantification than could be reached separately using the two techniques. In addition, its application allowed us to delineate a personalized urinary profile based on a list of compound levels covering a wide range of metabolic processes. Its expansion to more samples in the future will allow us to enlarge our knowledge of many metabolites' normal and abnormal values in human urine. Its translation into clinical practice can be of great value, such as identifying biomarkers of disease susceptibility and following the individual therapeutic outcomes [35]. These two aspects are among the main applications of metabolomics to improve the accuracy of personalized medicine.

Supplementary Materials: The following are available online. Table S1: List of the 164 metabolites identified and quantified by the SYNHMET approach; Figure S1: Personalized metabolic profile for subject 2852 showing the 164 metabolites identified and quantified by SYNHMET; Table S2: Comparison of the retention times observed for nine labeled standards with those assigned with the NMR–HRMS intensity correlation method in the two chromatographic conditions.

Author Contributions: Conceptualization, G.P. and D.O.C.; methodology, G.P. and D.O.C.; validation, G.P., G.C. and D.V.; formal analysis, G.P.; investigation, G.P. and D.O.C.; resources, E.M. and V.S.; data curation, G.P., S.L., C.M., R.S., A.S., F.M., R.V. and L.O.; writing—original draft preparation, D.O.C.; writing—review and editing, G.P., E.M. and D.O.C.; funding acquisition, V.S. All authors have read and agreed to the published version of the manuscript.

Funding: This research was funded by Collezione Nazionale di Composti Chimici e Centro Screening (CNCCS) Consortium, Grant: Project B, Sp 2, WP2, 2019 "Metabolomics Studies".

Institutional Review Board Statement: All the studies carried out on patients' samples were conducted according to the guidelines of the Declaration of Helsinki and approved by the Institutional Ethical Committee (IRCCS Ospedale San Raffaele, Milan, protocol URINEBIOMAR, approval date 14 July 2016).

Informed Consent Statement: Informed consent was obtained from all subjects involved in the study.

Data Availability Statement: Not applicable.

Conflicts of Interest: The authors declare no conflict of interest.

Sample Availability: Samples are not available from the authors.

References

1. German, J.B.; Hammock, B.D.; Watkins, S.M. Metabolomics: Building on a century of biochemistry to guide human health. *Metabolomics* **2005**, *1*, 3–9. [CrossRef] [PubMed]
2. Bory, C.; Boulieu, R.; Chantin, C.; Mathieu, M. Diagnosis of alcaptonuria: Rapid analysis of homogentisic acid by HPLC. *Clin. Chim. Acta.* **1990**, *189*, 7–11. [CrossRef]
3. Wishart, D.S. Metabolomics for investigating physiological and pathophysiological processes. *Physiol. Rev.* **2019**, *99*, 1819–1875. [CrossRef] [PubMed]
4. Bhinderwala, F.; Wase, N.; DiRusso, C.; Powers, R. Combining Mass Spectrometry and NMR Improves Metabolite Detection and Annotation. *J. Proteome Res.* **2018**, *17*, 4017–4022. [CrossRef]
5. Wishart, D.S. Emerging applications of metabolomics in drug discovery and precision medicine. *Nat. Rev. Drug Discov.* **2016**, *15*, 473–484. [CrossRef]
6. Xiao, J.F.; Zhou, B.; Ressom, H.W. Metabolite identification and quantitation in LC-MS/MS-based metabolomics. *TrAC—Trends Anal. Chem.* **2012**, *32*, 1–14. [CrossRef]
7. Marshall, D.D.; Lei, S.; Worley, B.; Huang, Y.; Garcia-Garcia, A.; Franco, R.; Dodds, E.D.; Powers, R. Combining DI-ESI–MS and NMR datasets for metabolic profiling. *Metabolomics* **2015**, *11*, 391–402. [CrossRef]
8. Marshall, D.D.; Powers, R. Beyond the paradigm: Combining mass spectrometry and nuclear magnetic resonance for metabolomics. *Prog. Nucl. Magn. Reson. Spectrosc.* **2017**, *100*, 1–16. [CrossRef]

9. Cassiède, M.; Nair, S.; Dueck, M.; Mino, J.; McKay, R.; Mercier, P.; Quémerais, B.; Lacy, P. Assessment of 1H NMR-based metabolomics analysis for normalization of urinary metals against creatinine. *Clin. Chim. Acta.* **2017**, *464*, 37–43. [CrossRef]

10. Teul, J.; Deja, S.; Celińska-Janowicz, K.; Ząbek, A.; Młynarz, P.; Barć, P.; Junka, A.; Smutnicka, D.; Bartoszewicz, M.; Pałka, J.; et al. LC-QTOF-MS and 1H NMR Metabolomics Verifies Potential Use of Greater Omentum for Klebsiella pneumoniae Biofilm Eradication in Rats. *Pathogens.* **2020**, *9*, 399. [CrossRef]

11. Hanifa, M.A.; Skott, M.; Maltesen, R.G.; Rasmussen, B.S.; Nielsen, S.; Frøkiær, J.; Ring, T.; Wimmer, R. Tissue, urine and serum NMR metabolomics dataset from a 5/6 nephrectomy rat model of chronic kidney disease. *Data Br.* **2020**, *33*. [CrossRef]

12. Trimigno, A.; Khakimov, B.; Savorani, F.; Poulsen, S.K.; Astrup, A.; Dragsted, L.O.; Engelsen, S.B. Human urine ^{1}H NMR metabolomics reveals alterations of the protein and carbohydrate metabolism when comparing habitual Average Danish diet vs. healthy New Nordic diet. *Nutrition* **2020**, *79–80*. [CrossRef]

13. Deng, L.; Fu, D.; Zhu, L.; Huang, J.; Ling, Y.; Cai, Z. Testosterone deficiency accelerates early stage atherosclerosis in miniature pigs fed a high-fat and high-cholesterol diet: Urine 1H NMR metabolomics targeted analysis. *Mol. Cell. Biochem.* **2020**, *476*, 1245–1255. [CrossRef]

14. Alinaghi, M.; Nguyen, D.N.; Bertram, H.C.; Sangild, P.T. Direct implementation of intestinal permeability test in nmr metabolomics for simultaneous biomarker discovery—a feasibility study in a preterm piglet model. *Metabolites* **2020**, *10*, 22. [CrossRef]

15. Gu, H.; Pan, Z.; Xi, B.; Hainline, B.E.; Shanaiah, N.; Asiago, V.; Gowda, G.A.N.; Raftery, D. ^{1}H NMR metabolomics study of age profiling in children. *NMR Biomed.* **2009**. [CrossRef]

16. Bouatra, S.; Aziat, F.; Mandal, R.; Guo, A.C.; Wilson, M.R.; Knox, C.; Bjorndahl, T.C.; Krishnamurthy, R.; Saleem, F.; Liu, P.; et al. The Human Urine Metabolome. *PLoS ONE* **2013**, *8*. [CrossRef]

17. Nicholson, J.K.; Connelly, J.; Lindon, J.C.; Holmes, E.; Kell, D.B.; Fiehn, O.; Weckwerth, W.; Plumb, R.S. Statistical Heterospectroscopy, an Approach to the Integrated Analysis of NMR and UPLC-MS Data Sets: Application in Metabonomic Toxicology Studies. *Nat. Rev. Drug Discov.* **2004**, *9*, 363–371. [CrossRef]

18. Zhou, W.; Yang, S.; Wang, P.G. Matrix effects and application of matrix effect factor. *Bioanalysis* **2017**, *9*, 1839–1844. [CrossRef]

19. Bales, J.R.; Higham, D.P.; Howe, I.; Nicholson, J.K.; Sadler, P.J. Use of high-resolution proton nuclear magnetic resonance spectroscopy for rapid multi-component analysis of urine. *Clin. Chem.* **1984**, *30*, 426–432. [CrossRef]

20. Bairaktari, E.; Katopodis, K.; Siamopoulos, K.C.; Tsolas, O. Paraquat-induced renal injury studied by 1H nuclear magnetic resonance spectroscopy of urine. *Clin. Chem.* **1998**, *44*, 1256–1261. [CrossRef]

21. Warrack, B.M.; Hnatyshyn, S.; Ott, K.-H.; Reily, M.D.; Sanders, M.; Zhang, H.; Drexler, D.M. Normalization strategies for metabonomic analysis of urine samples. *J. Chromatogr. B* **2009**, *877*, 547–552. [CrossRef] [PubMed]

22. Garde, A.H.; Hansen, M.; Kristiansen, J.; Knudsen, L.E. Comparison of Uncertainties Related to Standardization of Urine Samples with Volume and Creatinine Concentration. *Ann. Occup. Hyg.* **2004**, *48*, 171–179. [CrossRef] [PubMed]

23. Miller, R.C.; Brindle, E.; Holman, D.J.; Shofer, J.; Klein, N.A.; Soules, M.R.; O'Connor, K.A. Comparison of Specific Gravity and Creatinine for Normalizing Urinary Reproductive Hormone Concentrations. *Clin. Chem.* **2004**, *50*, 924–932. [CrossRef] [PubMed]

24. Cocker, J.; Mason, H.J.; Warren, N.D.; Cotton, R.J. Creatinine adjustment of biological monitoring results. *Occup. Med.* **2011**, *61*, 349–353. [CrossRef]

25. World Health Organization Biological Monitoring of Chemical Exposure in the Workplace: Guidelines. Available online: https://apps.who.int/iris/bitstream/handle/10665/41856/WHO_HPR_OCH_96.1.pdf;jsessionid=31E4ACD0F4E0FF269122167B8C89C255?sequence=1 (accessed on 14 June 2021).

26. Dinges, S.S.; Hohm, A.; Vandergrift, L.A.; Nowak, J.; Habbel, P.; Kaltashov, I.A.; Cheng, L.L. Cancer metabolomic markers in urine: Evidence, techniques and recommendations. *Nat. Rev. Urol.* **2019**, *16*, 339–362. [CrossRef]

27. Dona, A.C.; Kyriakides, M.; Scott, F.; Shephard, E.A.; Varshavi, D.; Veselkov, K.; Everett, J.R. A guide to the identification of metabolites in NMR-based metabonomics/metabolomics experiments. *Comput. Struct. Biotechnol. J.* **2016**, *14*, 135–153. [CrossRef]

28. Hrydziuszko, O.; Viant, M.R. Missing values in mass spectrometry based metabolomics: An undervalued step in the data processing pipeline. *Metabolomics* **2012**, *8*, 161–174. [CrossRef]

29. Wittmann, B.M.; Stirdivant, S.M.; Mitchell, M.W.; Wulff, J.E.; McDunn, J.E.; Li, Z.; Dennis-Barrie, A.; Neri, B.P.; Milburn, M.V.; Lotan, Y.; et al. Bladder cancer biomarker discovery using global metabolomic profiling of urine. *PLoS ONE* **2014**, *9*. [CrossRef]

30. A Community-Built Calibration System._ The Case Study of Quantification of Metabolites in Grape Juice by qNMR Spectroscopy | Elsevier Enhanced Reader. Available online: https://reader.elsevier.com/reader/sd/pii/S0039914020301466?token=498612BB72768CFBAFAB5F1CBEB76F7FE71A646F8304AF18A6E6004E28A988B6AB22E44F8F1F6035B6F24C6C9DB984D9&originRegion=eu-west-1&originCreation=20210701143417 (accessed on 1 July 2021).

31. Levey, A.S.; Inker, L.A.; Coresh, J. GFR estimation: From physiology to public health. *Am. J. Kidney Dis.* **2014**, *63*, 820–834. [CrossRef]

32. Nabity, M.B. Traditional Renal Biomarkers and New Approaches to Diagnostics. *Toxicol. Pathol.* **2018**, *46*, 999–1001. [CrossRef]

33. Adedeji, A.O.; Pourmohamad, T.; Chen, Y.; Burkey, J.; Betts, C.J.; Bickerton, S.J.; Sonee, M.; McDuffie, J.E. Investigating the Value of Urine Volume, Creatinine, and Cystatin C for Urinary Biomarkers Normalization for Drug Development Studies. *Int. J. Toxicol.* **2019**, *38*, 12–22. [CrossRef]

34. Wishart, D.S.; Feunang, Y.D.; Marcu, A.; Guo, A.C.; Liang, K.; Vázquez-Fresno, R.; Sajed, T.; Johnson, D.; Li, C.; Karu, N.; et al. HMDB 4.0: The human metabolome database for 2018. *Nucleic Acids Res.* **2018**, *46*, D608–D617. [CrossRef]

35. Li, B.; He, X.; Jia, W.; Li, H. Novel Applications of Metabolomics in Personalized Medicine: A Mini-Review. *Molecules* **2017**, *22*, 1173. [CrossRef]
36. Kouznetsova, V.L.; Kim, E.; Romm, E.L.; Zhu, A.; Tsigelny, I.F. Recognition of early and late stages of bladder cancer using metabolites and machine learning. *Metabolomics* **2019**, *15*, 94. [CrossRef]
37. Rybak, M.E.; Sternberg, M.R.; Pao, C.I.; Ahluwalia, N.; Pfeiffer, C.M. Urine excretion of caffeine and select caffeine metabolites is common in the US population and associated with caffeine intake. *J. Nutr.* **2015**, *145*, 766–774. [CrossRef]
38. Sinués, B.; Sáenz, M.A.; Lanuza, J.; Bernal, M.L.; Fanlo, A.; Juste, J.L.; Mayayo, E. Five caffeine metabolite ratios to measure tobacco-induced CYP1A2 activity and their relationships with urinary mutagenicity and urine flow. *Cancer Epidemiol. Biomarkers Prev.* **1999**, *8*, 159–166.
39. Cumberbatch, M.G.K.; Jubber, I.; Black, P.C.; Esperto, F.; Figueroa, J.D.; Kamat, A.M.; Kiemeney, L.; Lotan, Y.; Pang, K.; Silverman, D.T.; et al. Epidemiology of Bladder Cancer: A Systematic Review and Contemporary Update of Risk Factors in 2018. *Eur. Urol.* **2018**, *74*, 784–795. [CrossRef]
40. Ussher, J.R.; Elmariah, S.; Gerszten, R.E.; Dyck, J.R.B. The Emerging Role of Metabolomics in the Diagnosis and Prognosis of Cardiovascular Disease. *J. Am. Coll. Cardiol.* **2016**, *68*, 2850–2870. [CrossRef]
41. Xiao, C.; Hao, F.; Qin, X.; Wang, Y.; Tang, H. An optimized buffer system for NMR-based urinary metabonomics with effective pH control, chemical shift consistency and dilution minimization. *Analyst* **2009**, *134*, 916. [CrossRef]
42. Beckonert, O.; Keun, H.C.; Ebbels, T.M.D.; Bundy, J.; Holmes, E.; Lindon, J.C.; Nicholson, J.K. Metabolic profiling, metabolomic and metabonomic procedures for NMR spectroscopy of urine, plasma, serum and tissue extracts. *Nat. Protocols* **2007**, *2*, 2692–2703. [CrossRef]
43. Mckay, R.T. How the 1D-NOESY suppresses solvent signal in metabonomics NMR spectroscopy: An examination of the pulse sequence components and evolution. *Concepts Magn. Reson. Part A Bridg. Educ. Res.* **2011**, *38 A*, 197–220. [CrossRef]

molecules

Article

Glycosylation States on Intact Proteins Determined by NMR Spectroscopy

Audra A. Hargett [1], Aaron M. Marcella [1], Huifeng Yu [1], Chao Li [2], Jared Orwenyo [2], Marcos D. Battistel [1], Lai-Xi Wang [2] and Darón I. Freedberg [1,*]

1   Center for Biologics Evaluation and Review, Laboratory of Bacterial Polysaccharides, Food and Drug Administration (FDA), Silver Spring, MD 20993, USA; Audra.Hargett@fda.hhs.gov (A.A.H.); ammarcella.7@gmail.com (A.M.M.); huifeng.yu@fda.hhs.gov (H.Y.); marcos.battistel@fda.hhs.gov (M.D.B.)
2   Department of Chemistry and Biochemistry, University of Maryland, College Park, MD 20742, USA; chaoli@umd.edu (C.L.); nyabutoo@gmail.com (J.O.); wang518@umd.edu (L.-X.W.)
*   Correspondence: daron.freedberg@fda.hhs.gov

Abstract: Protein glycosylation is important in many organisms for proper protein folding, signaling, cell adhesion, protein-protein interactions, and immune responses. Thus, effectively determining the extent of glycosylation in glycoprotein therapeutics is crucial. Up to now, characterizing protein glycosylation has been carried out mostly by liquid chromatography mass spectrometry (LC-MS), which requires careful sample processing, e.g., glycan removal or protein digestion and glycopeptide enrichment. Herein, we introduce an NMR-based method to better characterize intact glycoproteins in natural abundance. This non-destructive method relies on exploiting differences in nuclear relaxation to suppress the NMR signals of the protein while maintaining glycan signals. Using RNase B Man5 and RNase B Man9, we establish reference spectra that can be used to determine the different glycoforms present in heterogeneously glycosylated commercial RNase B.

Keywords: glycosylated proteins; heteronuclear NMR; HSQC-TOCSY; natural abundance; T_2 filter; glycoprotein

Citation: Hargett, A.A.; Marcella, A.M.; Yu, H.; Li, C.; Orwenyo, J.; Battistel, M.D.; Wang, L.-X.; Freedberg, D.I. Glycosylation States on Intact Proteins Determined by NMR Spectroscopy. *Molecules* **2021**, *26*, 4308. https://doi.org/10.3390/molecules26144308

Academic Editor: Robert Brinson

Received: 9 June 2021
Accepted: 6 July 2021
Published: 16 July 2021

Publisher's Note: MDPI stays neutral with regard to jurisdictional claims in published maps and institutional affiliations.

1. Introduction

Glycosylation is one of the most common post-translational modifications (PTM). There are two main types of glycosylation: (i) *O*-linked glycosylation, in which glycans are covalently linked to the hydroxyl oxygen of serine (S) or threonine (T) residues [1,2], and (ii) *N*-linked glycosylation, where glycans are attached to asparagine (N) residues within the N-X-S/T sequon [3–5]. In *N*-linked glycosylation, the initial glycan moiety, $Glc_3Man_9GlcNAc_2$, is transferred to the nascent polypeptide chain co-translationally in the ER, and then the initial glycan is processed in the ER and Golgi apparatus resulting in either a high-mannose, hybrid, or complex type N-glycan (Figure S1). Because protein glycosylation is not template driven, it is inherently heterogeneous, with several factors contributing to the final glycan structure, such as protein structure [6,7], enzyme protein levels [8], Golgi transport mechanism [9], and secretory protein load [10]. Overall, this process yields heterogeneously glycosylated proteins, such as IgG, which has 32 possible glycans for its one *N*-linked glycosylation site at N297 [11].

For many glycoproteins, the glycans are critical to the protein's structure, stability, and function [12–14]. For example, monoclonal antibodies (mAbs) that lack core fucose in the Fc region (remote to the antigen binding site) lead to an increase in antibody-dependent cell-mediated cytotoxicity [15–18]. IgG sialylation has been linked to anti-inflammatory activity [19]. The loss of some HIV-1 gp120 glycans leads to an increase in protein degradation and a decrease in binding to the host cell receptor [20–22]. In Hepatitis C virus envelope 2 protein, the loss of either N2 or N4 glycan results in total loss of HCV infectivity [23]. These are just a few glycoproteins where the location and type of glycan are critical to protein function. Thus, developing tools to characterize intact glycoproteins will aid in the understanding of optimal glycosylation for a given function, especially in protein therapeutics.

To improve our understanding of structure/function and to ensure proper glycosylation of protein therapeutics, the glycans must be fully characterized. Typically, mass spectrometry (MS)-based methods are combined with other methods, such as glycan enrichment, affinity separation, enzymatic digestion, liquid chromatography (LC) and/or gas chromatography (GC), to determine protein glycosylation [24–26]. However, the stereochemistry of a glycan, including the type of glycosidic linkage, are challenging to determine by MS, because it is difficult to distinguish between isobaric species like glucose (Glc), galactose (Gal), and mannose (Man). To overcome these limitations, a direct, robust and simple NMR spectroscopy method was recently proposed for the detection and identification of protein glycoforms by denaturing the glycoprotein in urea [24]. This method provides a significant advantage by indirectly detecting modifications on intact proteins without sophisticated sample preparation or isotopic labeling. Moreover, the method is not limited by the protein's molecular weight due to the more favorable nuclear relaxation properties of denatured proteins. RNAse A and RNase B have identical amino acid sequences, but RNase B is glycosylated. In this report, we show that the glycans in intact folded RNase B can be characterized by NMR spectroscopy.

As a proof of concept, we chose RNase B glycoprotein as a model system because it is characterized by the following key properties: it is a ~15 kDa glycoprotein, a size that is amenable to NMR and enables the study of native glycosylation; it contains a single glycosylation site at N34 yet, it exists as five glycosylated variants (Man$_{5-9}$GlcNAc$_2$) and therefore, RNase B permits the study of the potential microheterogeneity in a single glycosylation (at N34) [27]; finally, previous studies of RNase B can be used to cross-validate our findings. In a study of commercial RNase B, the oligosaccharides were released and isolated, and the relative molar portions of Man$_5$ to Man$_9$ were determined to be 57, 31, 4, 7, and 1%, respectively [28].

In pioneering work, Brown showed that differential T_2s can be used to distinguish between fluids with different viscosities [29]. Herein, we build on this idea, using ^{1}H-^{13}C HSQC-TOCSY [30,31], with varied mixing times on natural abundance samples for fast detection and analysis of glycoprotein microheterogeneity, without complicated sample preparation. The mixing time efficiently relaxes away protein resonances and, although this phenomenon is not unexpected, it hasn't been investigated in detail [32]. In this report, we show that using a T_2 filter in small glycoproteins reduces the spectral complexity that arises from the protein peaks yet captures the glycosylation microheterogeneity by retaining glycan peaks.

2. Results and Discussion

2.1. ^{1}H-^{15}N HSQC of RNase A and RNase B

^{1}H-^{15}N HSQC spectra of unlabeled RNase A and RNase B were collected at 700 MHz, 37 °C in 5 h with all expected signals, consistent with previous results [33]. Peaks were assigned based on ^{1}H-^{15}N chemical shifts deposited in the Biological Magnetic Resonance Data Bank (BMRB) for RNase A. Like mapping protein ligand binding sites by comparing apo and bound forms' ^{1}H-^{15}N chemical shift changes, protein backbone amino acid chemical shifts can be affected by PTMs. Backbone resonance assignments of RNase A/B provide useful data that were used to identify the effect of glycosylation on the polypeptide chain. Upon glycosylation, the backbone ^{1}H-^{15}N chemical shift perturbation in RNase B compared to RNase A is confined to the region around the glycosylation site (±4 amino acids, Figure S2). Minimal changes were observed for most of the glycoprotein's NMR signals. However, measurable differences were observed at N34 (glycosylation site). Specifically, T36 shifts 0.064 ppm in ^{1}H, and S32, N34, and K37 change by 0.014 ppm in ^{1}H. In ^{15}N, S32, N34, T36, and K37 change by 0.28, 0.93, 0.63, and 0.28 ppm, respectively. Interestingly, R33 is absent in the RNase B spectrum, and L35's chemical shift is unchanged. Thus, in RNase

B only polar or charged residues proximal to the glycosylation site exhibit a change in chemical shift. It may also be that both charged residues and N34, the glycosylated residue, are exposed, thus when N34 is glycosylated, other exposed residues are affected. While these ^{1}H-^{15}N spectra suggest that PTMs effect the protein and the location of attachment, they do not provide accurate information regarding the precise identity of the modification.

2.2. ^{1}H-^{13}C HSQC of RNase B Man5 and RNase B Man9

In contrast to natural abundance ^{1}H-^{15}N HSQC, natural abundance ^{1}H-^{13}C HSQC spectra are higher in sensitivity and can provide substantially more information regarding protein glycosylation. The larger number of ^{13}C atoms in an amino acid than ^{15}N atoms increases spectral complexity; nevertheless, the uniqueness of ^{13}C chemical shift ranges and NMR experiments can be used to differentiate between protein and glycan subspectra.

Glycan anomeric ^{1}H-^{13}C correlations occur in a unique spectral region which does not overlap with most protein signals [24,34], since both nuclei are typically deshielded in ^{1}H (4.3–5.8 ppm) and ^{13}C (98–106 ppm). In the case of pure glycans or single glycoforms, the number of anomeric peaks can be used to determine the number of saccharide residues in a given glycan. ^{1}H-^{13}C HSQC spectra were taken of two engineered RNase B glycoproteins (Figures 1 and 2), each uniformly glycosylated at N34 with either Man$_5$GlcNAc$_2$ (Man$_5$) or Man$_9$GlcNAc$_2$ (Man$_9$). Intact electrospray ionization mass spectrometry (ESI-MS) showed that each glycoprotein contained one predominant mass after charge state deconvolution (14898 Da, RNase B Man$_5$; 15546 Da, RNase B Man$_9$, Figure 3). Figure S1 provides schematics of different types of glycan and their linkages. For RNase B Man$_5$, seven anomeric peaks were unambiguously observed (Figure S3a). GlcNAc2, Man3, Man4, Man4′, ManA, and ManB have ^{13}C chemical shifts between 100 and 144 ppm and ^{1}H chemical shifts between 4.6 and 5.2 ppm. The GlcNAc$_1$ anomeric peak is shifted significantly, in ^{13}C, to 78.4 ppm as it is amide linked to the protein [24]. Notably, there are no overlapping protein chemical shifts in this region.

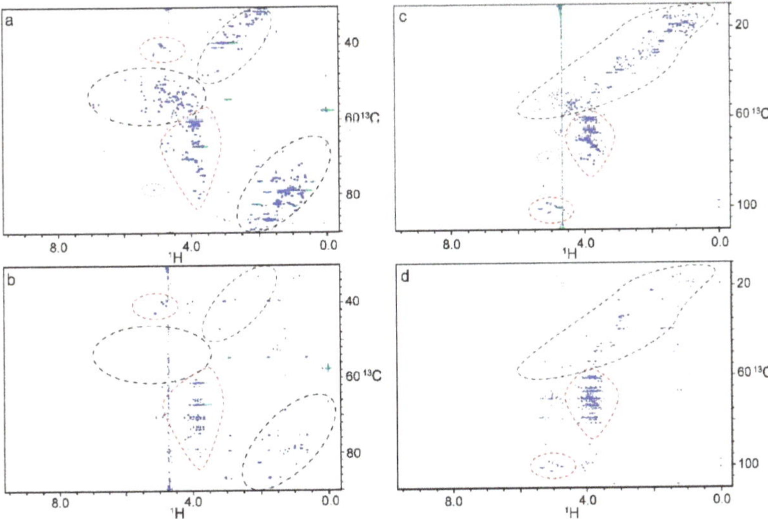

Figure 1. Comparison of ^{1}H-^{13}C HSQC and HSQC-TOCSY spectra of RNase B Man$_5$ and Man$_9$. (**a**) HSQC spectrum of 0.3 mM RNase B Man$_5$ contains peaks for both protein (black circles) and glycans (red circles). RNase B Man$_5$ anomeric protons (upper red circle) are folded in the ^{13}C dimension and range from 98–103 ppm. (**b**) An HSQC-TOCSY with a 90 ms mixing time of RNase B Man$_5$ takes advantage of the longer glycan T$_2$, so that the glycans peaks are retained while the protein peaks are greatly reduced. Similarly, an HSQC of 0.6 mM RNase B Man$_9$ (**c**) shows peaks for both protein and glycan, while in the HSQC-TOCSY experiment (**d**) mostly glycan peaks are retained.

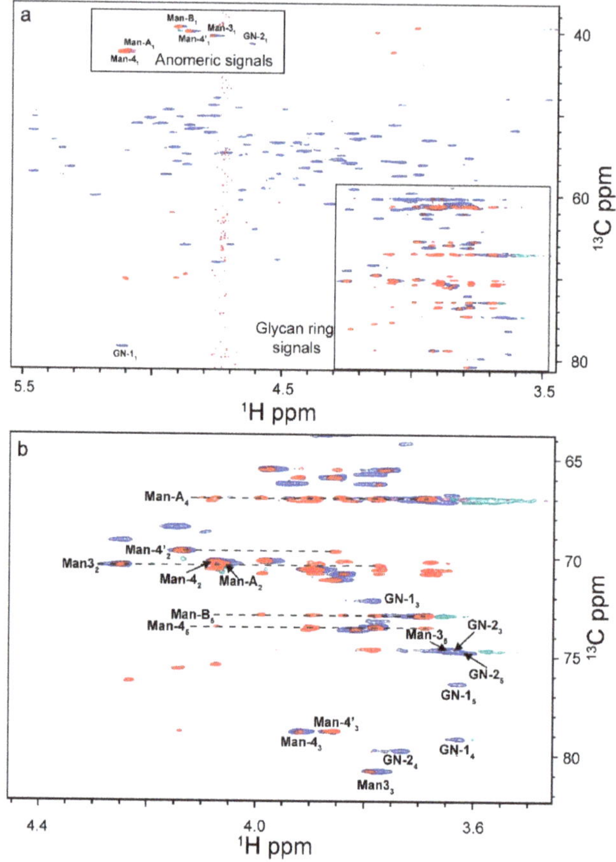

Figure 2. Overlay of a 2D ^{1}H-^{13}C HSQC (blue) and ^{1}H-^{13}C HSQC-TOCSY (red) of 0.3 mM RNase B Man$_5$ at pH 6, 37 °C (blue) and ^{1}H-^{13}C HSQC-TOCSY (red) (**a**). Chemical shifts of ^{13}C anomeric signals are folded and range from 98 to 104 ppm, except GlcNAc$_1$ which is more shielded with a ^{13}C chemical shift of 78 ppm. (**b**) Glycan ring signals with lines drawn to show the 90 ms TOCSY correlations for each of the monosaccharides except GlcNac$_1$ (GN1) and GlcNAc$_2$ (GN2).

In a ^{1}H-^{13}C HSQC of RNase B Man$_9$, anomeric correlations are observed in a similar spectral region as RNase B Man$_5$ (Figure S3b). However, because RNase B Man$_9$ contains additional Man residues D1-D3, with α1-2 linkages, their signals overlap and were resolved with Lorentz-to-Gauss processing for line narrowing [35]. RNase B Man$_9$ provides an additional challenge because the NMR signals for the C2-C6 positions on each glycan significantly overlap. Although the spectra were collected at 34 Hz/pt ^{13}C resolution, it is still insufficient to resolve the individual signals within the ring. Nevertheless, the unique fingerprint in the anomeric region is the ideal method for distinguishing glycoforms. The most evident signals are those belonging to ManA (+D3), ManC (+D1), ManB (+D2) and Man4 (+C). These three signals are unique to Man$_9$, as Man$_5$ does not contain a 'C' residue and ManA and Man4 are more deshielded in the ^{1}H dimension when linked to the D mannoses. A similar strategy was used to characterize the glycoprofile of FcεRiα [36]. These researchers were able to assess the different glycoforms using HSQC spectra of uniformly ^{15}N/^{13}C-labeled glycoproteins under both folded and denatured sample conditions at lower concentrations than we report. They also report assessing the relative abundances of each glycoform using the anomeric region only. Thus, both their methods and those we report here can be used to assess glycoforms.

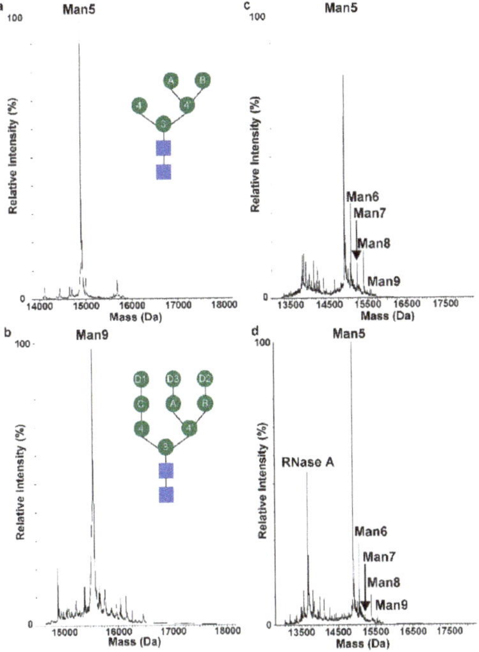

Figure 3. Deconvoluted ESI-MS spectra of (**a**) RNase B Man5, (**b**) RNase B Man9, (**c**) vendor 1 RNase B, and (**d**) vendor 2 RNase B. RNase B Man5 and RNase B Man9 are singly glycosylated. Both vendor 1 and vendor 2 contain a population of high-mannose glycan with predominantly GlcNAc2Man5. The vendor 2 sample did have a significant population of RNase A that was not present in Vendor 1's sample.

2.3. Relaxation Selection for Glycan Regions of Spectrum

To enrich glycan regions of the spectrum for peak assignment and reduce ambiguity observed in ring regions of the spectrum, ^{1}H-^{13}C HSQC-TOCSY experiments were used [30,31]. TOCSY mixing times were optimized by a simple linewidth analysis. Data collected at a sufficient resolution to obtain reliable linewidths in both ^{1}H and ^{13}C dimensions can be used to estimate the upper limits of the T_2s, using the relation $(T_2 \approx \left(\pi * \left(\Delta \nu_{\frac{1}{2}} \right) \right)^{-1}$, where $\Delta \nu_{\frac{1}{2}}$ is the linewidth at half height [37]. Table S1 shows a list of peaks corresponding mostly to either the ring or anomeric region of the N-glycan or ^{1}H, ^{13}Cα peaks from the protein. Based on linewidths, the range of ^{13}C transverse relaxation times for the protein specific regions is 9–13 ms with an average of 11.6 ms, whereas in the glycan regions the T_2 range is 12 to 17 ms and an average of 14.4 ms. This yields an approximate difference in relaxation time of 25% between the glycan and protein components, limiting the amount of relaxation effect to exploit. In contrast to the ^{13}C relaxation times, ^{1}H relaxation times displayed a greater disparity between the protein and glycan resonances. The protein-specific relaxation times in ^{1}H were between 10 and 35 ms, with an average of 16.4 ms. The glycan relaxation times in ^{1}H ranged from 14 to 45 ms and averaged 29 ms. This provides a nearly twofold (80%) difference in relaxation times which is easier and more effective to exploit. The average T_2 determined from this analysis was then used to plot transverse magnetization loss over time (Figure S4). This allows for quantitatively selecting mixing times to maximize the intensity difference between the protein and glycan peaks. Because the relaxation rate difference is nearly twofold, it allows most of the protein signals to relax while maintaining enough glycan signal so as not to increase experiment time.

Signal-to-noise ratios (SNR) in protein dominant regions (2a/2b) and glycan dominant regions (1a/1b) were assessed in an HSQC and HSQC-TOCSY of RNase B Man$_5$ (Table 1, Figure S5). The glycan regions maintain 43.3% of their signal intensity in the HSQC-TOCSY (90 ms mixing time), compared to the HSQC, where in the protein regions only an average of 11.8% of the initial intensity remains. This 3.7-fold difference agrees with the estimated signal loss calculated using the relaxation times (3.2-fold) and significantly simplifies the spectra while also providing the benefit of intra-ring correlations of coupled ^{1}Hs through the TOCSY (Figures 1 and 2). Interestingly, signal loss is observed for glycan residues GlcNAc1 and GlcNac2 which are spatially close to the protein and have a T$_2$ closer to that of the protein C$_\alpha$. Other NMR experiments such as the HSQC-ROESY have a similar effect on protein signal attenuation.

Table 1. Peak volume between glycan dominant (1a/1b) and protein dominant (2a/2b) spectral regions.

Region	^{1}H (ppm)	^{13}C (ppm)	HSQC Peak Volume	HSQC-TOCSY Peak Volume	% Remaining Volume
1a	4.5-5.5	37.5-45.0	1416.69	584.12	41.2
1b	3.5-4.5	60.0-80.0	14461.11	6563.13	45.4
2a	3.5-5.5	45.0-60.0	13203.96	883.33	6.7
2b	0.2-3.5	30.0-90.0	42549.51	7195.17	16.9

2.4. Analysis of Commercial RNase B Samples

Using the uniformly glycosylated RNase glycoproteins as references, two commercially available RNase B samples were evaluated. RNase B from vendor 1 was reported to be 80% pure, and RNase B from vendor 2 was reported to be 50% pure. All RNase B samples were analyzed for glycosylation heterogeneity and purity using ESI-MS. Mass spectra of intact RNase B were collected and a charge envelope consisting of +8 to +15 charged ions were observed for each of the samples. The charge state envelope was deconvoluted [38] using the Waters MassLynx MS software and the glycosylation pattern was determined for each of the RNase B samples (Figure 3). The commercial RNase B from vendor 1 contained predominantly GlcNAc$_2$Man$_5$ at N34 (exp = 14,898 Da, calc = 14,897 Da), with a small percentage of GlcNAc$_2$Man$_{6-9}$. Similarly, commercial RNase B from vendor 2 was mostly glycosylated with GlcNAc$_2$Man$_5$; however, this sample also contained RNase A (exp = 13,682 Da, calc = 13,681 Da). To have similar amounts of RNase B for the NMR analysis in both vendors, the percent of RNase A was accounted for when determining RNase B sample concentration for vendor 2. Overall, the distribution of N-glycans in RNase B is similar between the two manufacturers which should lead to nearly identical samples in the NMR experiments.

Initial ^{1}H-^{13}C HSQC analysis of the commercial RNase B revealed a contaminating peak present in vendor 1's sample (Figure S6, blue) that was not observed in the ESI-MS analysis (data not shown). It is possible that this glycoside-like molecule is a methyl mannoside that was either not completely removed after lectin affinity chromatography or was used to stabilize RNase B. Due to similarities in chemical shift between this contaminant and the RNase B glycan and its high SNR in the HSQC-TOCSY, vendor 1 RNase B was dialyzed using a 1 kDa MWCO membrane to remove the contaminant (Figure S6, red). After the dialysis, there were only minor differences between the vendor RNase B samples (Figure 4). Specifically, RNase B from vendor 2 contained peaks from 2.5–3.0 ppm ^{1}H and 30–40 ppm ^{13}C that are not present in vendor 1's RNase B. The chemical shifts that correspond to the glycan anomeric (4.5–5.5 ppm ^{1}H and 95–105 ppm ^{13}C) and ring regions (3.3–4.3 ppm ^{1}H and 60–80 ppm ^{13}C) are the same in both vendor RNase B spectra.

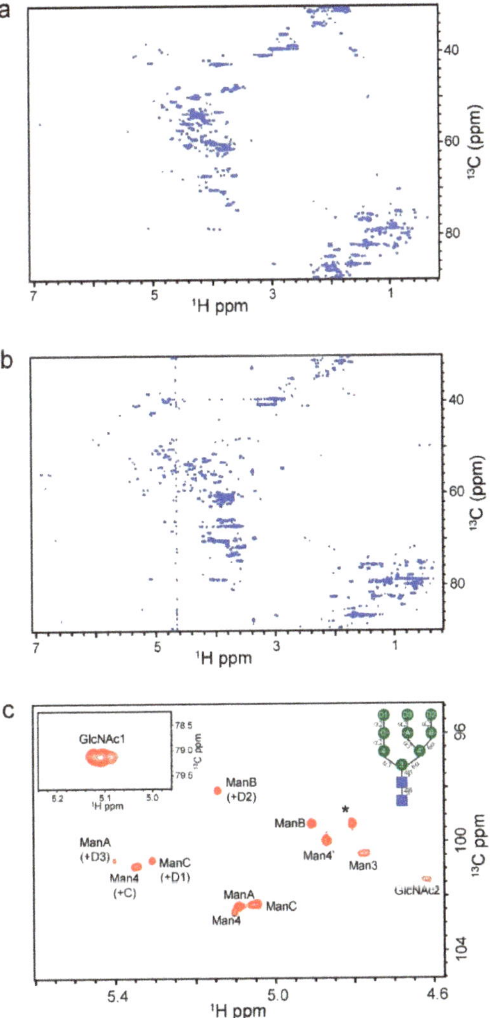

Figure 4. 2D ^{1}H-^{13}C HSQC of (**a**) 1.4 mM vendor 1 RNase B and (**b**) 1.0 mM vendor 2 RNase B. Glycan anomeric region and ring region are the same between the two vendors; however, the vendor 2 spectrum contains additional peaks from 2.5 to 3.0 ppm in ^{1}H and 30 to 40 ppm in ^{13}C that are not present in the RNase B vendor 1 spectrum. (**c**) Vendor 1 glycan anomeric signals used for quantitative analysis (inset: schematic of Man$_9$ glycan).

The ^{1}H-^{13}C HSQC and ^{1}H-^{13}C HSQC-TOCSY spectra from the vendor samples, which contained a heterogenous population of glycans (Man$_{5-9}$GlcNAc$_2$), was compared to the uniformly glycosylated reference RNase B spectra. Figure 4 shows the anomeric region of the commercially available RNase B from vendor 1 with transferred assignments from literature values [33]. In addition to the peaks observed in the RNase B Man$_5$ reference spectrum, there are additional peaks corresponding to Man 4, A, B, and C as each of these positions can be further modified by an α1-2 linked mannose residue. Man D1, D2, and D3 chemical shifts overlap Man A, C, and 4 and cannot be assigned at the current spectral resolution. Thus, the ratios of the entire glycan population cannot be qualitatively estimated using these signals. Nevertheless, the SNR of some of the glycan anomeric signals can be used for quantification, as we show below.

2.5. Quantitative Analysis of Commercial RNase B Glycoforms

To normalize the results for quantitative analysis, all experiments performed were collected on a 700 MHz (^{1}H) magnet equipped with cryoprobe, which provided increased sensitivity. This is especially useful in experiments carried out at natural abundance, as performed in the present study. One HSQC was run with nearly identical experimental conditions for the three samples analyzed (RNase B Man$_5$, RNase B Man$_9$, and commercial RNase B (vendor 1)), protein concentrations were between 18–22 mg/mL. The temperature was set to 25 °C for Man$_5$ and commercial RNase B and 37 °C for Man$_9$ RNase B. In all cases, the lowest SNR was observed for GlcNAc1 and GlcNAc2 anomeric signals. The SNRs of GlcNAc1 were standardized to account for differences in protein concentration between samples, as all experiments were performed with the same number of scans and t_1 points. RNase B Man$_9$ GlcNAc1 had a SNR of 19:1, RNase B Man$_5$ GlcNAc1 had a s/n of 16:1 and commercial RNase B 16:1.

Quantitative ratios of each glycoform present in the commercial RNase B are difficult to obtain due to differential T_2 relaxation. For example, Man C and Man 3 cannot be compared, since Man 3 is far more restricted and would be expected to have more efficient relaxation, leading to a decrease in peak intensity that would reduce accuracy in assessing relative abundance of glycoforms. Therefore, to better estimate the relative glycoforms abundance, only residues with similar T_2s can be compared such as Man B and Man C. The ratio of Man B: Man C correlates to glycoforms GlcNAc$_2$Man$_5$ and GlcNAc$_2$Man$_6$. This ratio was determined to be 1.8:1 by peak height comparison, which is in line with the reported estimate of 1.84:1 [28]. Another ratio that should be close to 1:1 is that of ManC (+D1):ManB (+D2) as there is more GlcNAc$_2$Man$_8$ than GlcNAc$_2$Man$_7$ according to the MS analysis leading to equal amounts of the two terminal mannose residues present in GlcNAc$_2$Man$_8$. In this sample the ratio was 1.04:1 in line with the expectation.

3. Materials and Methods

RNase B from bovine pancreas (Cat. R7884) purchased from Sigma-Aldrich St. Louis, MO, USA) and VWR and all other chemicals were purchased from Sigma-Aldrich (unless otherwise noted). RNase A from bovine pancreas were from Roche (Cat. 10109142001).

3.1. Preparation of RNase B with Homogenous Man$_5$ and Man$_9$ N-Glycans

The synthesis of RNase B Man$_9$ followed our previously reported method [39]. ESI-MS: theoretical mass for RNase B Man$_9$, M = 15,546 Da; found (deconvolution data) (m/z) 15,547 Da. RNase B Man$_5$ was prepared following a similar chemoenzymatic method. Briefly, the Man$_5$ oxazoline was obtained by α1,2-mannosidease catalyzed hydrolysis of Man$_9$ N-glycan, followed by Endo-A treatment to provide the Man$_5$GlcNAc, which was converted to Man5GlcNAc-oxazoline by treatment with 2-chloro-1,3-dimethylimidazolium chloride (DMC) and triethylamine in water [40]. A solution of Man$_5$GlcNAc-oxazoline (500 μg, 0.49 μmol) and GlcNAc-RNase (500 μg, 0.036 μmol) was incubated with EndoA-N171A (200 μg) in buffer (PBS, 100 mM, pH 7.4, 10 μL) at 30 °C for 8 h. The reaction was monitored by analytical HPLC, and the glycoprotein product was isolated by preparative HPLC to give Man$_5$GlcNAc$_2$-RNase as a white foam after lyophilization (418 μg, 78%). ESI-MS: calc'd. for RNase B Man$_5$, M = 14,897 Da; found (deconvolution data) (m/z) 14,898 Da.

Analytical reverse-phase high-performance liquid chromatography (RP-HPLC) was performed on a Waters 626 HPLC instrument equipped with an YMC-Triart C18 column (5 μm, 4.6 × 250 mm) for reverse phase. The YMC-Triart column was eluted using a linear gradient of acetonitrile (22–29%, v/v) with water containing 0.1% TFA over 35 min at a flowrate of 0.5 mL/min under UV 280. The LC-ESI-MS was performed on an Exactive™ Plus Orbitrap mass spectrometer (Thermo Scientific) equipped with a C8 column (Poroshell

300SB-C8, 1.0 × 75 mm, 5 μm, Agilent). Mass spectra were analyzed, and deconvolution of MS data was obtained by MagTran.

3.2. NMR of Glycoproteins

All NMR experiments were performed at 700 MHz ^{1}H Frequency (16.5 T magnet) with a Bruker Avance III HD console and 5 mm triple gradient TCI cryoprobe. RNase B solution concentrations varied from 5-20 mg/mL, depending on sample, in 20 mM phosphate buffer at pH* 6.0 with added DSS as an internal reference in ~99% D_2O. For HSQC (Bruker pulse sequence hsqcetgpsi) and HSQC-TOCSY (Bruker pulse sequence hsqcdietgpsi) experiments, 4096 and 768 total points in ^{1}H and ^{13}C, respectively, were collected. Non-uniform sampling was used in the indirect dimension with 30–50% of the points collected based on the schedules from the Wagner group [41]. Data were collected with spectral windows of 7002.801 Hz (10 ppm) and 10,563.504 Hz (60 ppm), with carrier frequencies of 4.7 ppm and 60 ppm in ^{1}H and ^{13}C, respectively, and were reconstructed using SMILE [42]. The anomeric ^{13}C peaks appear at ~40 ppm, due to folding in most of the spectra collected. Data were processed with zero-filling to 2x the total points collected in ^{1}H and ^{13}C. A square cosine bell window function was applied in both dimensions. For linewidth measurements, data were collected with traditional sampling and 2048 × 1024 total points in ^{1}H and ^{13}C, respectively. Spectral widths of 10 ppm (7002.8 Hz) in ^{1}H and 100 ppm (17,605.8 Hz) in ^{13}C were used with the ^{13}C carrier set at 60 ppm (10,562.9 Hz). Data were processed in the same way as described above.

For ^{1}H-^{15}N experiments, samples of RNase B and RNase A were dissolved in 20 mM phosphate buffer pH 6.5 in 2.5% D_2O at ~15 mg/mL (~1 mM). ^{1}H-^{15}N experiments were collected using an HSQC pulse sequence with a flip back pulse and WATERGATE [43] element for water suppression. Acquisition times of 41 and 26 ms for ^{1}H and ^{15}N respectively were used with spectral resolution of 12 and 19 Hz/pt. High-quality data were collected over ~10 h of experiment time; two 10 h spectra were added to increase signal to noise.

ESI-MS of RNase B: 2.5 μL of a 0.5 mg/mL solution of RNAse B was injected into a C18 trap column and eluted using a gradient from 0 to 40% acetonitrile in acetate buffer, pH 4.5, at a flow rate of 0.5 μL/min. Data were collected on a Waters Synapt G2 HDMS system with a nanoAcquity LC system. Data containing the entire charge state envelope were deconvoluted using the Masslynx software yielding the mass of the singly charged species.

MALDI-TOF MS of RNase B: 0.5 mg/mL RNase B was mixed 1:1 with dihydroxybenzoic acid matrix solution (10 mg/mL in 50:50 acetonitrile:water with 0.1 % TFA). The mixture was then spotted on a stainless steel MALDI target and allowed to air dry. Samples were analyzed using a Bruker Autoflex Speed instrument with a voltage of 13 keV, 4000 shots per spectrum and delay time of 800 ns. Samples were all shot in positive ion mode with singly, doubly, and triply charged states observed.

4. Conclusions

Multiple RNase B samples were tested using a standard set of HSQC and HSQC-TOCSY pulse sequences with varying mixing times. The size of RNase B offers a protein with favorable T_2 relaxation times when compared to even larger proteins and even more beneficial is the expected mobility of the N-glycan. The RNase B N-glycan was not observed in the crystal structure suggesting an unrestrained conformation which allowed us to exploit the relaxation differences between the protein and glycan. This difference in protein and carbohydrate relaxation times provides the opportunity to analyze the two components of the spectra independently on increasingly complex samples. We will use these experiments to fine-tune the conditions under which NMR spectra of polysaccharide conjugate vaccines can be better analyzed.

Supplementary Materials: The following are available online, Figure S1: Schematic of N-glycans using symbol nomenclature; Figure S2: Overlay of 1H-15N HSQC of RNase A and RNase B; Figure S3: Anomeric region of the 1H-13C HSQC of uniformly glycosylated RNase B Man5 and RNase B-Man9;

Figure S4: Plots of the average T2 relaxation for proteins and glycans; Figure S5: Intensity regions used in Table 1 for the 1H-13C HSQC and the HSQC-TOCSY for RNase B Man5. Table S1: Linewidths and estimated T2 relaxation values for protein Cα's and glycan anomeric ring 1H-13C correlations.

Author Contributions: A.A.H.—ran experiments and contributed to manuscript writing; A.M.M.—ran experiments and contributed to manuscript writing; H.Y.—ran experiments; C.L.—prepared glycosylated RNase B Man$_5$ and RNase B-Man$_9$; J.O.—prepared glycosylated RNase B Man$_5$ and RNase B-Man$_9$; M.D.B.—Trained and assisted in NMR data acquisition and processing; L.-X.W.—manuscript preparation, supervision and funding; D.I.F.—Project design and implementation, manuscript preparation, supervision and funding. All authors have read and agreed to the published version of the manuscript.

Funding: NIH Grant R01GM080374 to L.X.W.

Institutional Review Board Statement: Not applicable.

Informed Consent Statement: Not applicable.

Data Availability Statement: Not applicable.

Conflicts of Interest: The authors declare no conflict of interest.

Sample Availability: Not applicable.

References

1. Schachter, H. The clinical relevance of glycobiology. *J. Clin. Invest.* **2001**, *108*, 1579–1582. [CrossRef]
2. Trimble, R.B.; Lubowski, C.; Hauer, C.R., 3rd; Stack, R.; McNaughton, L.; Gemmill, T.R.; Kumar, S.A. Characterization of N- and O-linked glycosylation of recombinant human bile salt-stimulated lipase secreted by Pichia pastoris. *Glycobiology* **2004**, *14*, 265–274. [CrossRef] [PubMed]
3. Kornfeld, R.; Kornfeld, S. Assembly of asparagine-linked oligosaccharides. *Annu. Rev. Biochem.* **1985**, *54*, 631–664. [CrossRef]
4. Robbins, P.W.; Hubbard, S.C.; Turco, S.J.; Wirth, D.F. Proposal for a common oligosaccharide intermediate in the synthesis of membrane glycoproteins. *Cell* **1977**, *12*, 893–900. [CrossRef]
5. Rothman, J.E.; Lodish, H.F. Synchronised transmembrane insertion and glycosylation of a nascent membrane protein. *Nature* **1977**, *269*, 775–780. [CrossRef]
6. Hang, I.; Lin, C.W.; Grant, O.C.; Fleurkens, S.; Villiger, T.K.; Soos, M.; Morbidelli, M.; Woods, R.J.; Gauss, R.; Aebi, M. Analysis of site-specific N-glycan remodeling in the endoplasmic reticulum and the Golgi. *Glycobiology* **2015**, *25*, 1335–1349. [CrossRef]
7. Suga, A.; Nagae, M.; Yamaguchi, Y. Analysis of protein landscapes around N-glycosylation sites from the PDB repository for understanding the structural basis of N-glycoprotein processing and maturation. *Glycobiology* **2018**, *28*, 774–785. [CrossRef]
8. Oka, T.; Ungar, D.; Hughson, F.M.; Krieger, M. The COG and COPI complexes interact to control the abundance of GEARs, a subset of Golgi integral membrane proteins. *Mol. Biol. Cell* **2004**, *15*, 2423–2435. [CrossRef]
9. Hossler, P.; Mulukutla, B.C.; Hu, W.S. Systems analysis of N-glycan processing in mammalian cells. *PLoS ONE* **2007**, *2*, e713. [CrossRef] [PubMed]
10. Jimenez-Mallebrera, C.; Torelli, S.; Feng, L.; Kim, J.; Godfrey, C.; Clement, E.; Mein, R.; Abbs, S.; Brown, S.C.; Campbell, K.P.; et al. A comparative study of alpha-dystroglycan glycosylation in dystroglycanopathies suggests that the hypoglycosylation of alpha-dystroglycan does not consistently correlate with clinical severity. *Brain Pathol.* **2009**, *19*, 596–611. [CrossRef]
11. Arnold, J.N.; Wormald, M.R.; Sim, R.B.; Rudd, P.M.; Dwek, R.A. The impact of glycosylation on the biological function and structure of human immunoglobulins. *Annu. Rev. Immunol* **2007**, *25*, 21–50. [CrossRef] [PubMed]
12. Varki, A. Biological roles of oligosaccharides: All of the theories are correct. *Glycobiology* **1993**, *3*, 97–130. [CrossRef]
13. Lowe, J.B.; Marth, J.D. A genetic approach to Mammalian glycan function. *Annu. Rev. Biochem.* **2003**, *72*, 643–691. [CrossRef]
14. Moremen, K.W.; Tiemeyer, M.; Nairn, A.V. Vertebrate protein glycosylation: Diversity, synthesis and function. *Nat. Rev. Mol. Cell Biol.* **2012**, *13*, 448–462. [CrossRef]
15. Davies, J.; Jiang, L.; Pan, L.Z.; LaBarre, M.J.; Anderson, D.; Reff, M. Expression of GnTIII in a recombinant anti-CD20 CHO production cell line: Expression of antibodies with altered glycoforms leads to an increase in ADCC through higher affinity for FC gamma RIII. *Biotechnol. Bioeng.* **2001**, *74*, 288–294. [CrossRef]
16. Shields, R.L.; Lai, J.; Keck, R.; O'Connell, L.Y.; Hong, K.; Meng, Y.G.; Weikert, S.H.; Presta, L.G. Lack of fucose on human IgG1 N-linked oligosaccharide improves binding to human Fcgamma RIII and antibody-dependent cellular toxicity. *J. Biol. Chem.* **2002**, *277*, 26733–26740. [CrossRef]
17. Shinkawa, T.; Nakamura, K.; Yamane, N.; Shoji-Hosaka, E.; Kanda, Y.; Sakurada, M.; Uchida, K.; Anazawa, H.; Satoh, M.; Yamasaki, M.; et al. The absence of fucose but not the presence of galactose or bisecting N-acetylglucosamine of human IgG1 complex-type oligosaccharides shows the critical role of enhancing antibody-dependent cellular cytotoxicity. *J. Biol. Chem.* **2003**, *278*, 3466–3473. [CrossRef]

18. Umana, P.; Jean-Mairet, J.; Moudry, R.; Amstutz, H.; Bailey, J.E. Engineered glycoforms of an antineuroblastoma IgG1 with optimized antibody-dependent cellular cytotoxic activity. *Nat. Biotechnol.* **1999**, *17*, 176–180. [CrossRef]

19. Washburn, N.; Schwab, I.; Ortiz, D.; Bhatnagar, N.; Lansing, J.C.; Medeiros, A.; Tyler, S.; Mekala, D.; Cochran, E.; Sarvaiya, H.; et al. Controlled tetra-Fc sialylation of IVIg results in a drug candidate with consistent enhanced anti-inflammatory activity. *Proc. Natl. Acad. Sci. USA* **2015**, *112*, E1297–E1306. [CrossRef] [PubMed]

20. Kong, L.; Wilson, I.A.; Kwong, P.D. Crystal structure of a fully glycosylated HIV-1 gp120 core reveals a stabilizing role for the glycan at Asn262. *Proteins* **2015**, *83*, 590–596. [CrossRef] [PubMed]

21. Mathys, L.; Francois, K.O.; Quandte, M.; Braakman, I.; Balzarini, J. Deletion of the highly conserved N-glycan at Asn260 of HIV-1 gp120 affects folding and lysosomal degradation of gp120, and results in loss of viral infectivity. *PLoS ONE* **2014**, *9*, e101181. [CrossRef]

22. Wei, Q.; Hargett, A.A.; Knoppova, B.; Duverger, A.; Rawi, R.; Shen, C.H.; Farney, S.K.; Hall, S.; Brown, R.; Keele, B.F.; et al. Glycan Positioning Impacts HIV-1 Env Glycan-Shield Density, Function, and Recognition by Antibodies. *iScience* **2020**, *23*, 101711. [CrossRef] [PubMed]

23. Goffard, A.; Callens, N.; Bartosch, B.; Wychowski, C.; Cosset, F.L.; Montpellier, C.; Dubuisson, J. Role of N-linked glycans in the functions of hepatitis C virus envelope glycoproteins. *J. Virol.* **2005**, *79*, 8400–8409. [CrossRef] [PubMed]

24. Schubert, M.; Walczak, M.J.; Aebi, M.; Wider, G. Posttranslational modifications of intact proteins detected by NMR spectroscopy: Application to glycosylation. *Angewandte Chemie* **2015**, *127*, 7202–7206. [CrossRef]

25. Marino, K.; Bones, J.; Kattla, J.J.; Rudd, P.M. A systematic approach to protein glycosylation analysis: A path through the maze. *Nat. Chem. Biol.* **2010**, *6*, 713–723. [CrossRef]

26. Karamanos, N.K.; Hjerpe, A. Strategies for analysis and structure characterization of glycans/proteoglycans by capillary electrophoresis. Their diagnostic and biopharmaceutical importance. *J. Chromatogr. B Biomed. Appl.* **1999**, *13*, 507–512. [CrossRef]

27. Dwek, R.A.; Edge, C.J.; Harvey, D.J.; Wormald, M.R.; Parekh, R.B. Analysis of glycoprotein-associated oligosaccharides. *Annu. Rev. Biochem.* **1993**, *62*, 65–100. [CrossRef]

28. Fu, D.; Chen, L.; O'Neill, R.A. A detailed structural characterization of ribonuclease B oligosaccharides by 1H NMR spectroscopy and mass spectrometry. *Carbohydr. Res.* **1994**, *261*, 173–186. [CrossRef]

29. Brown, R.J.S. Proton Relaxation in Crude Oils. *Nature* **1961**, *189*, 387–388. [CrossRef]

30. Lerner, L.; Bax, A. Sensitivity-Enhanced Two-Dimensional Heteronuclear Relayed Coherence Transfer NMR-Spectroscopy. *J. Magn. Reson.* **1986**, *69*, 375–380. [CrossRef]

31. Lerner, L.; Bax, A. Application of new, high-sensitivity ^{1}H-^{13}C-N.M.R.- spectral techniques to the study of oligosaccharides. *Carbohydr. Res.* **1987**, *166*, 35–46. [CrossRef]

32. Wyss, D.F.; Dayie, K.T.; Wagner, G. The counterreceptor binding site of human CD2 exhibits an extended surface patch with multiple conformations fluctuating with millisecond to microsecond motions. *Protein. Sci.* **1997**, *6*, 534–542. [CrossRef] [PubMed]

33. Blanchard, V.; Frank, M.; Leeflang, B.R.; Boelens, R.; Kamerling, J.P. The structural basis of the difference in sensitivity for PNGase F in the de-N-glycosylation of the native bovine pancreatic ribonucleases B and BS. *Biochemistry* **2008**, *47*, 3435–3446. [CrossRef]

34. Wütrich, K. *NMR of Proteins and Nucleic Acids*, 1st ed.; John Wiley and Sons: New York, NY, USA, 1986.

35. Shahzad-Ul-Hussan, S.; Sastry, M.; Lemmin, T.; Soto, C.; Loesgen, S.; Scott, D.A.; Davison, J.R.; Lohith, K.; O'Connor, R.; Kwong, P.D.; et al. Insights from NMR Spectroscopy into the Conformational Properties of Man-9 and Its Recognition by Two HIV Binding Proteins. *Chembiochem* **2017**, *18*, 764–771. [CrossRef]

36. Unione, L.; Lenza, M.P.; Arda, A.; Urquiza, P.; Lain, A.; Falcon-Perez, J.M.; Jimenez-Barbero, J.; Millet, O. Glycoprofile Analysis of an Intact Glycoprotein As Inferred by NMR Spectroscopy. *ACS Cent. Sci.* **2019**, *5*, 1554–1561. [CrossRef]

37. Becker, E.D. *High Resolution NMR Spectroscopy*, 3rd ed.; Springer Science & Business Media: New York, NY, USA, 2001.

38. Ferrige, A.G.; Seddon, M.J.; Jarvis, S.; Skilling, J.; Aplin, R. Maximum entropy deconvolution in electrospray mass spectrometry. *Rapid Commun. Mass Spectrom.* **1991**, *5*, 374–377. [CrossRef]

39. Huang, W.; Li, C.; Li, B.; Umekawa, M.; Yamamoto, K.; Zhang, X.; Wang, L.X. Glycosynthases enable a highly efficient chemoenzymatic synthesis of N-glycoproteins carrying intact natural N-glycans. *J. Am. Chem. Soc.* **2009**, *131*, 2214–2223. [CrossRef] [PubMed]

40. Amin, M.N.; McLellan, J.S.; Huang, W.; Orwenyo, J.; Burton, D.R.; Koff, W.C.; Kwong, P.D.; Wang, L.X. Synthetic glycopeptides reveal the glycan specificity of HIV-neutralizing antibodies. *Nat. Chem. Biol.* **2013**, *9*, 521–526. [CrossRef]

41. Hyberts, S.G.; Milbradt, A.G.; Wagner, A.B.; Arthanari, H.; Wagner, G. Application of iterative soft thresholding for fast reconstruction of NMR data non-uniformly sampled with multidimensional Poisson Gap scheduling. *J. Biomol. NMR* **2012**, *52*, 315–327. [CrossRef]

42. Ying, J.; Delaglio, F.; Torchia, D.A.; Bax, A. Sparse multidimensional iterative lineshape-enhanced (SMILE) reconstruction of both non-uniformly sampled and conventional NMR data. *J. Biomol. NMR* **2017**, *68*, 101–118. [CrossRef] [PubMed]

43. Piotto, M.; Saudek, V.; Sklenar, V. Gradient-tailored excitation for single-quantum NMR spectroscopy of aqueous solutions. *J. Biomol. NMR* **1992**, *2*, 661–665. [CrossRef]

molecules

MDPI

Article

Use of the 2D ^{1}H-^{13}C HSQC NMR Methyl Region to Evaluate the Higher Order Structural Integrity of Biopharmaceuticals

Tsang-Lin Hwang [†], **Dipanwita Batabyal** [†], **Nicholas Knutson** and **Mats Wikström** *

Attribute Sciences, Amgen Inc., Thousand Oaks, CA 91320, USA; thwang@amgen.com (T.-L.H.);
dbatabya@amgen.com (D.B.); nsk2@g.uclas.edu (N.K.)
* Correspondence: matsw@amgen.com
† Equal contributions.

Abstract: The higher-order structure (HOS) of protein therapeutics is directly related to the function and represents a critical quality attribute. Currently, the HOS of protein therapeutics is characterized by methods with low to medium structural resolution, such as Fourier transform infrared (FTIR), circular dichroism (CD), intrinsic fluorescence spectroscopy (FLD), and differential scanning calorimetry (DSC). High-resolution nuclear magnetic resonance (NMR) methods have now been introduced, representing powerful approaches for HOS characterization (HOS by NMR). NMR is a multi-attribute method with unique abilities to give information on all structural levels of proteins in solution. In this study, we have compared 2D ^{1}H-^{13}C HSQC NMR with two established biophysical methods, i.e., near-ultraviolet circular dichroism (NUV-CD) and intrinsic fluorescence spectroscopy, for the HOS assessments for the folded and unfolded states of two monoclonal antibodies belonging to the subclasses IgG1 and IgG2. The study shows that the methyl region of the ^{1}H-^{13}C HSQC NMR spectrum is sensitive to both the secondary and tertiary structure of proteins and therefore represents a powerful tool in assessing the overall higher-order structural integrity of biopharmaceutical molecules.

Keywords: higher-order structure; tertiary structure; fluorescence; circular dichroism; NMR; HOS by NMR; product characterization; biopharmaceuticals

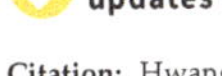

Citation: Hwang, T.-L.; Batabyal, D.; Knutson, N.; Wikström, M. Use of the 2D ^{1}H-^{13}C HSQC NMR Methyl Region to Evaluate the Higher Order Structural Integrity of Biopharmaceuticals. *Molecules* **2021**, *26*, 2714. https://doi.org/10.3390/molecules26092714

Academic Editor:
Thomas Mavromoustakos

Received: 29 March 2021
Accepted: 29 April 2021
Published: 5 May 2021

Publisher's Note: MDPI stays neutral with regard to jurisdictional claims in published maps and institutional affiliations.

1. Introduction

The higher-order structure (HOS) of proteins includes the secondary, tertiary, and quaternary structure, and represents a critical quality attribute directly related to the structural integrity and the function of therapeutic proteins. The characterization of HOS represents a significant challenge for biopharmaceuticals and is currently being performed using low- to medium-resolution biophysical methods, such as Fourier transform infrared spectroscopy (FTIR), circular dichroism (CD) spectroscopy, intrinsic fluorescence spectroscopy (FLD), and differential scanning calorimetry (DSC) [1,2]. With the increasing interest in different protein modalities in biopharmaceutical development and the rapidly expanding area of biosimilar development, there is a growing need for new analytical methods with higher specificity than the methods commonly applied. During the development and lifecycle of protein therapeutics, the innovator product will most often go through multiple process changes, in which it is required to show that any process-related drug product variations are within the acceptable criteria, and therefore considered comparable. In a similar fashion, it is required to show similarity between the biopharmaceutical reference product and developed biosimilars. The application of nuclear magnetic resonance (NMR) for the assessment of HOS has been suggested as a technology with the potential to more accurately assess differences in HOS as compared to established methods [3]. This technology, referred to as Profile NMR, is based on a one-dimensional diffusion NMR method, in which the strong signals from excipients are efficiently suppressed by dephasing the signals through gradients due to faster Brownian motions of smaller excipient molecules as compared to larger protein in the sample, leaving a spectrum of the protein product only [4,5]. In

129

addition to the 1D NMR method, a 2D ^{1}H-^{13}C HSQC method was introduced [6], which shows great promise for the HOS assessment of monoclonal antibodies (mAbs) [7]. Finally, mass spectrometric methods, such as hydrogen-deuterium exchange experiments, have also gained considerable interest for the assessment of biopharmaceuticals [8,9].

In this study, we have compared two established methods, near-ultraviolet circular dichroism (NUV CD) and intrinsic fluorescence (FLD) spectroscopy, for the assessment of HOS for biopharmaceuticals against a 2D ^{1}H-^{13}C HSQC NMR method modified to suppress signals from excipients. To demonstrate the effect HOS has on each spectroscopic method, we compared the folded and unfolded states of two monoclonal antibody subclasses, IgG1 and IgG2, with about 95% sequence identity.

2. Results

The NUV-CD spectra of the folded and unfolded states of IgG1 and IgG2 are shown in Figure 1. The effects of HOS on the differential absorption of left and right circularly polarized light can be seen in the spectral comparisons of the folded and unfolded states of IgG1 and IgG2, in Figure 1A,B, respectively. In general, the NUV-CD spectra of native proteins are characterized by distinct features at around 293 and 286 nm attributable to tryptophan, at 285 to 270 nm attributable to tyrosine and tryptophan, and 250–265 nm attributable to phenylalanine, superimposed over the disulfide signal from 250 to 280 nm. While the unfolded spectra of both IgG's show relatively featureless lines close to zero (Figure 1D), the folded spectra show absorption changes for the chromophores: tryptophan, tyrosine, and phenylalanine, indicating that these pendent groups are incorporated into highly organized portions of the protein, i.e., tertiary structure. Furthermore, even small differences in HOS and primary structure give rise to unique spectra for the folded states of the two mAbs, allowing them to be distinguished from each other as well (Figure 1C).

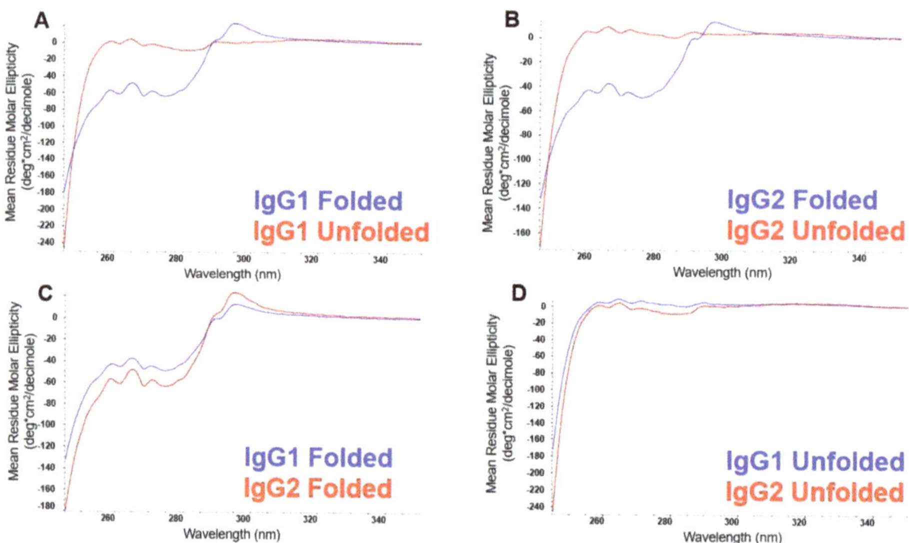

Figure 1. NUV-CD spectra of the folded and unfolded samples of IgG1 (**A**) and IgG2 (**B**). Comparison of the spectra from the folded states of the IgG1 and IgG2 molecules in (**C**), and the unfolded states for these two molecules in (**D**).

The FLD spectra of the folded and unfolded states of IgG1 and IgG2 are shown in Figure 2. The emission wavelengths of the internal fluorophores: tryptophan, phenylalanine, and tyrosine, are sensitive to the polarity of their environments. Higher polarity environments, particularly water from the solvent, cause the wavelengths of emission to lengthen (i.e., red shift). Therefore, unfolded proteins with more solvent-exposed fluorophores will appear more red-shifted than proteins whose tertiary structure tends to sequester these fluorophores in internal, more non-polar environments (Figure 2A,B) [10]. In our study, for both the mAbs IgG1 and IgG2, the folded spectra have a peak around 323 nm, and upon unfolding, the peak shifts to around 345 nm. We also observe that the fluorescence intensity upon unfolding increases (by almost 30%) for both mAbs and is due to the fact that the fluorescence quenching groups are further apart in the unfolded protein than in the native protein, resulting in significant lowering of energy transfer efficiency in the native protein [11]. However, little else about the HOS of the mAbs can be seen by their essentially indistinguishable folded spectra (Figure 2C). In addition, as clearly indicated in Figure 2D, the unfolded spectra for the two mAbs are essentially identical.

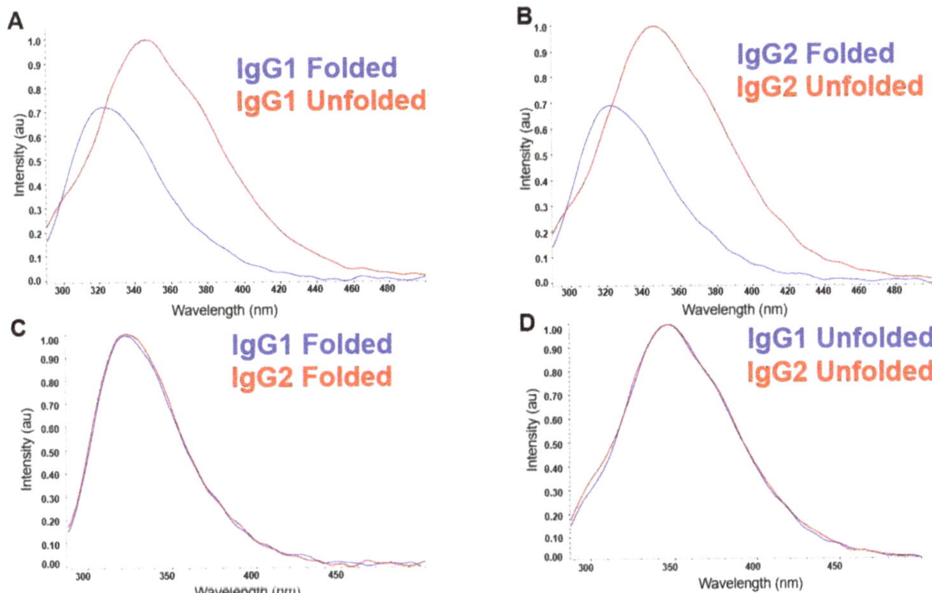

Figure 2. FLD spectra of the folded and unfolded samples of IgG1 (**A**) and IgG2 (**B**). Comparison of the spectra from the folded states of the IgG1 and IgG2 molecules in (**C**), and the unfolded states for these two molecules in (**D**).

The ^{1}H-^{13}C HSQC NMR methyl spectra of the folded and unfolded states of IgG1 and IgG2 are shown in Figure 3. NMR relays information about the local magnetic environments of the nuclei under investigation both through chemical bonds and spatially by the other atoms surrounding them. Atoms in more magnetically shielded environments have lower chemical shifts (plotted in ppms), while atoms in less shielded environments have higher chemical shifts. 2D ^{1}H-^{13}C HSQC NMR experiments are designed to correlate both protons (x-axis) and the carbon-13 atoms (y-axis) that they are directly attached to in a molecule. Since proteins are almost entirely composed of protons and carbons, NMR provides a wealth of information about primary structure and all levels of HOS. The resolution of primary structure can be seen in the unfolded spectra of IgG1 and IgG2 (Figure 3B,D). In the absence of any ordered secondary or tertiary structure, the unique chemical signals of different amino acids are resolved and by comparison to reported random-coil (unfolded) chemical shifts, side-chain units, particularly the methyl groups, can be tentatively assigned (Figure 4) [12]. In the folded state, the various magnetic environments of each individual amino acid disperse the side-chain signals to produce a truly unique spectrum for each protein, which is dependent upon all levels of HOS, as shown in Figure 3A,C.

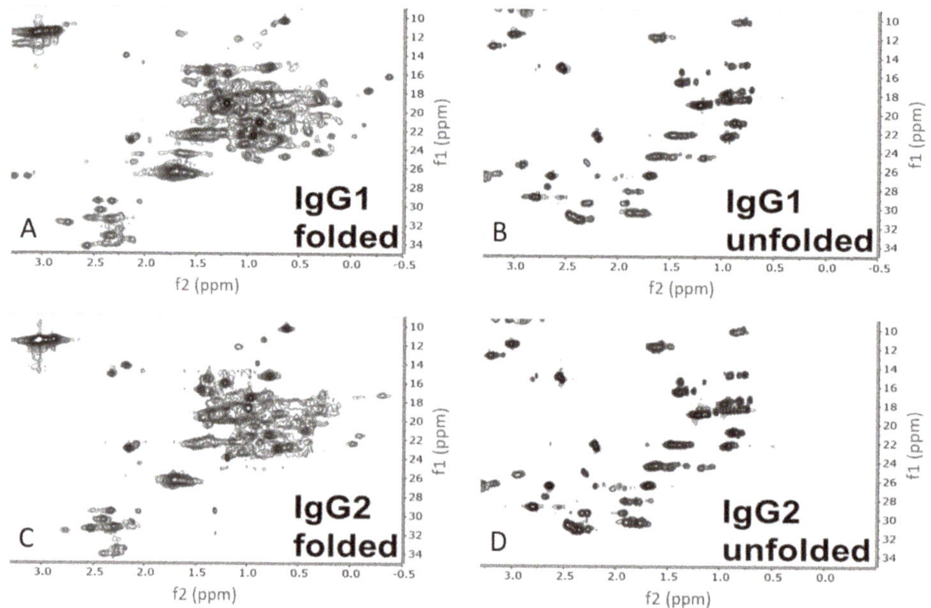

Figure 3. 2D ^{1}H-^{13}C HSQC spectra of the folded and unfolded samples of IgG1 and IgG2. (**A**,**B**) show IgG1 in the folded and unfolded state respectively, whereas (**C**,**D**) show IgG2 in the folded and unfolded states respectively. ^{1}H is represented on the x axis (f2), and ^{13}C is represented on the y axis (f1).

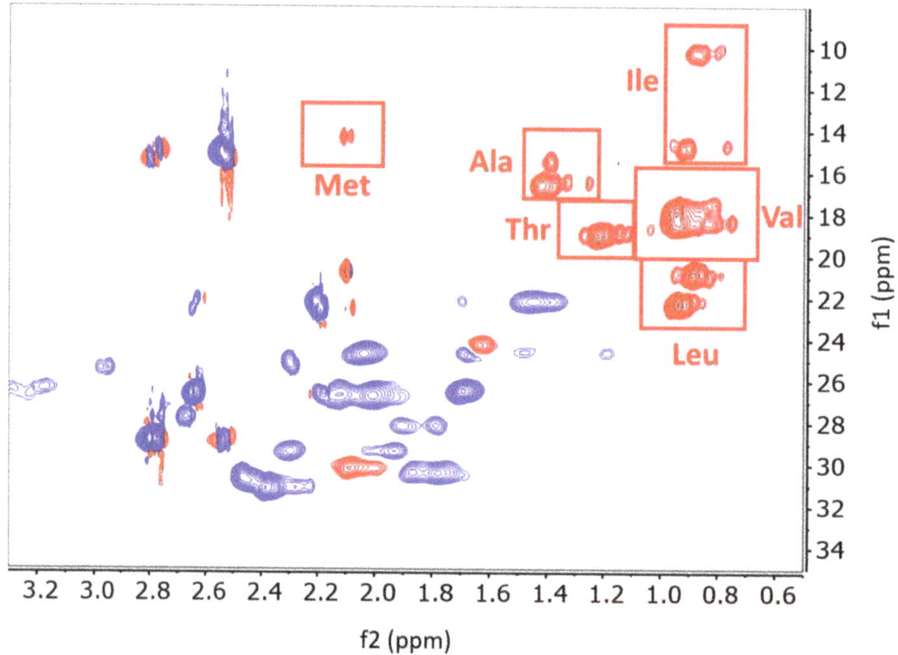

Figure 4. Multiplicity-edited 2D ^{1}H-^{13}C HSQC spectra of the unfolded state of IgG2. Red peaks are either CH or CH$_3$, blue peaks are CH$_2$. The methyl groups of each amino acid are shown in red boxes corresponding to the random coil ^{1}H and ^{13}C shift ranges for the six methyl groups [12]. ^{1}H is represented on the x axis (f2), and ^{13}C is represented on the y axis (f1).

3. Discussion

In this study, we have compared the ability of NUV-CD, FLD, and 2D NMR to measure HOS in two monoclonal antibody subclasses, IgG1 and IgG2. Unlike NUV-CD and FLD, which are only able to infer structural integrity from a limited number of chromophores in a protein, 2D NMR provides structural information about the entire molecule and is hence sensitive to even subtle changes in all levels of HOS. If low-resolution spectroscopic methods such as NUV-CD and FLD currently set the bar for assessing the structural integrity of biopharmaceuticals, we propose that vastly more informative 2D ^{1}H-^{13}C HSQC NMR methods become a replacement in many cases for this type of HOS assessments, and for the product characterization of biopharmaceuticals.

4. Materials and Methods

4.1. Sample Preparation

The test solutions were prepared from 100 mg/mL monoclonal antibodies IgG1 and IgG2 in the formulation buffer: 10 mM sodium acetate buffer, 9% (*w/v*) sucrose, at pH 5.2. The sequence identity comparing the IgG1 and Ig2 antibodies is 95% [13]. The IgG1 molecule harbors glycosylation on N302, while the IgG2 molecule contains glycosylation modifications on N298. Stock solutions of intact (folded) IgG1 and IgG2 were prepared at 50 mg/mL in the same formulation buffer. Stock solutions of the denatured (unfolded) IgG1 and IgG2 were prepared at 50 mg/mL in the formulation buffer, with 6M urea and 50 mM Tris (2-carboxyethyl) phosphine hydrochloride (TCEP).

4.2. Intrinsic Tryptophan Fluorescence Spectroscopy

Intrinsic tryptophan fluorescence spectra were obtained using an Applied Photophysics qCD Chirascan equipped with a fluorimeter at ambient temperature using cuvettes with a path length of 1 cm. Samples were run with an excitation wavelength of 280 nm, an excitation bandwidth of 5 nm, boxcar width of 5 nm, and averaged over 10 scans, with each scan taking 1 s. All mAb samples were diluted to approximately 0.033 mg/mL with buffer before measurements. Each sample was measured in triplicate and buffer blanks were subtracted before data analysis. The spectra were overlaid with each other and the similarity was compared by calculating the variability of the maximum fluorescence intensity and the wavelength at the maximum fluorescence.

4.3. Near-Ultraviolet Circular Dichroism Spectroscopy

The NUV-CD spectra were obtained on an Applied Photophysics qCD Chirascan spectropolarimeter at ambient temperature. The protein samples were analyzed at a concentration of about 0.5 mg/mL (both folded and unfolded). Using cuvettes with a pathlength of 1 cm, the spectra were corrected for concentration and contributions from the buffer and are reported as Mean Residue Molar Ellipticity. Each spectrum is an average of 4 scans and was smoothed with a 7-point smoothing function using the OMNIC 32 software (Thermo Fisher Scientific Inc.). Background nitrogen blanks and buffer blanks were measured to eliminate the signals from the nitrogen, cuvette, and buffer. The parameters for the Near UV CD were: 1 cm path length, 240–350 nm wave range, 2 s exposure time, 1 nm bandwidth, 0.5 nm step, and averaged over 4 runs with a 900 μL sample volume.

4.4. Nuclear Magnetic Resonance

A Bruker Avance III 600 MHz NMR spectrometer equipped with a 5 mm CPTCI cryoprobe was used to acquire NMR data at 310 K (37 °C) Bruker Biospin Corp, Billerica, MA, USA). Samples were prepared in 5 mm step-down NMR tubes (Wilmad LabGlass, Vineland, NJ, USA) with 5% D_2O. A modified 2D gradient-selected, sensitivity-enhanced ^{1}H-^{13}C HSQC NMR method [14] with additional excipient signal suppression was used to acquire the methyl fingerprints of the samples. The WET scheme [15] was used to suppress the acetate signal, and the asymmetric adiabatic pulse (HS1/2, R = 10, 0.9 Tp;

tanh/tan, R = 50, 0.1 Tp), with pulse length 375 µs [16], was applied to suppress the carbon signals of the sucrose while exciting the methyl ^{13}C signals of the protein. 2D ^{1}H-^{13}C HSQC experiments for Figure 3 used the following parameters to acquire NMR data: The f_2 spectral width was 14 ppm centered on 4.7 ppm with 2048 points. The f_1 spectral width was 28 ppm centered on 21 ppm. Spectra were acquired with 128 increments with 50% non-uniform sampling and 2048 scans in each increment, with recycle delay 0.5 s between scans. The total experimental time was 26.5 h for each spectrum. Digital filtering for 0.4 ppm bandwidth was used to further remove the water signal. GARP decoupling was applied during the WET scheme with 2.08 kHz RF power and the t2 acquisition with 4.16 kHz RF power. Shifted sine-squared bell window functions and zero filling were applied to both dimensions before Fourier transform of the data. The final spectra were 4 k × 1 k. The spectrum in Figure 4 was acquired using the ^{1}H-^{13}C multiplicity-edited HSQC (hsqcedetgpsisp2.2 in the Bruker library). The f_2 spectral width was 9 ppm, centered on 4.5 ppm with 2048 points. The f_1 spectral width was 160 ppm centered on 80 ppm. The 2D data were obtained with 512 increments and 16 scans in each increment, with recycle delay 1 s between scans. The total experimental time was 2.8 h. The data processing was carried out using the spectrometer software (TopSpin, Bruker BioSpin Corp, Billerica, MA, USA) and Mnova software (Mestrelab Research S.L., Santiago de Compostela, Spain).

5. Conclusions

The results show that 2D ^{1}H-^{13}C HSQC NMR is incredibly sensitive to primary, secondary, tertiary, and quaternary structures, and provides unique fingerprints for both the IgG1 and IgG2 subclasses used. Near-ultraviolet circular dichroism (NUV-CD) is also able to differentiate between the two IgG subclasses, while intrinsic fluorescence (FLD) is only able to distinguish between the folded and unfolded states of each protein, but not able to distinguish IgG1 from IgG2. When the 2D NMR methyl fingerprints are visually compared to the results from NUV-CD and FLD, the degree of HOS information captured by 2D NMR is vastly superior to that of either currently established method. Our findings therefore exemplify the superiority of NMR in the assessment of higher-order structural attributes of biopharmaceuticals.

Author Contributions: Conceptualization, M.W.; Formal analysis, D.B. and N.K.; Investigation, T.-L.H., D.B. and N.K.; Methodology, T.-L.H.; Resources, T.-L.H.; Supervision, M.W.; Writing—original draft, D.B. and N.K.; Writing—review and editing, M.W.; Manuscript reviewing, all authors. All authors have read and agreed to the published version of the manuscript.

Funding: This research received no external funding.

Acknowledgments: The authors would like to acknowledge the technical contributions of Jette Wypych.

Conflicts of Interest: The authors declare no conflict of interest.

Abbreviations

HOS	Higher-order structure
HOS by NMR	higher-order structure by nuclear magnetic resonance
FTIR	Fourier transform infrared
NUV-CD	near-ultraviolet circular dichroism
FLD	intrinsic fluorescence spectroscopy
NMR	nuclear magnetic resonance
ppm	parts per million
mAb	monoclonal antibody

References

1. Berkowitz, S.A.; Engen, J.R.; Mazzeo, J.R.; Jones, G.B. Analytical tools for characterizing biopharmaceuticals and the implications for biosimilars. Nature reviews. *Drug Discov.* **2012**, *11*, 527–540. [CrossRef] [PubMed]
2. Gabrielson, J.P.; Weiss, W.F., 4th. Technical Decision-Making with Higher Order Structure Data: Starting a New Dialogue. *J. Pharm. Sci.* **2015**, *104*, 1240–1245. [CrossRef] [PubMed]
3. Wen, J.; Batabyal, D.; Knutson, N.; Lord, H.; Wikström, M. A Comparison Between Emerging and Current Biophysical Methods for the Assessment of Higher-Order Structure of Biopharmaceuticals. *J. Pharm. Sci.* **2020**, *109*, 247–253. [CrossRef] [PubMed]
4. Poppe, L.; Jordan, J.B.; Lawson, K.; Jerums, M.; Apostol, I.; Schnier, P.D. Profiling Formulated Monoclonal Antibodies by ^{1}H NMR Spectroscopy. *Anal. Chem.* **2013**, *85*, 9623–9629. [CrossRef] [PubMed]
5. Poppe, L.; Jordan, J.B.; Rogers, G.; Schnier, P.D. On the Analytical Superiority of 1D NMR for Fingerprinting the Higher Order Structure of Protein Therapeutics Compared to Multidimensional NMR Methods. *Anal. Chem.* **2015**, *87*, 5539–5545. [CrossRef] [PubMed]
6. Arbogast, L.W.; Brinson, R.G.; Marino, J.P. Mapping monoclonal antibody structure by 2D ^{13}C NMR at natural abundance. *Anal. Chem.* **2015**, *87*, 3556–3561. [CrossRef] [PubMed]
7. Brinson, R.G.; Marino, J.P.; Delaglio, F.; Arbogast, L.W.; Evans, R.M.; Kearsley, A.; Gingras, G.; Ghasriani, H.; Aubin, Y.; Pierens, G.K.; et al. Enabling adoption of 2D-NMR for the higher order structure assessment of monoclonal antibody therapeutics. *mAbs* **2019**, *11*, 94–105. [CrossRef] [PubMed]
8. Zhang, Z.; Smith, D.L. Determination of amide hydrogen exchange by mass spectrometry: A new tool for protein structure elucidation. *Protein Sci.* **1993**, *2*, 522–531. [CrossRef] [PubMed]
9. Goswami, D.; Zhang, J.; Bondarenko, P.; Zhang, Z. MS-based conformation analysis of recombinant proteins in design, optimization and development of biopharmaceuticals. *Methods* **2018**, *144*, 134–151. [CrossRef] [PubMed]
10. Ramachander, R.; Jiang, Y.; Li, C.; Eris, T.; Young, M.; Dimitrova, M.; Narhi, L. Solid state fluorescence of lyophilized proteins. *Anal. Biochem.* **2008**, *376*, 173–182. [CrossRef] [PubMed]
11. Lakowicz, J.R. *Principles of Fluorescence Spectroscopy*, 2nd ed.; Springer: New York, NY, USA, 1999.
12. Wishart, D.S.; Bigam, C.G.; Holm, A.; Hodges, R.S.; Sykes, B.D. ^{1}H, ^{13}C and ^{15}N random coil NMR chemical shifts of the common amino acids. I. Investigations of nearest-neighbor effects. *J. Biomol. NMR* **1995**, *5*, 67–81. [CrossRef] [PubMed]
13. Dillon, T.M.; Ricci, M.S.; Vezina, C.; Flynn, G.C.; Liu, Y.D.; Rehder, D.S.; Plant, M.; Henkle, B.; Li, Y.; Deechongkit, S.; et al. Structural and Functional Characterization of Disulfide Isoforms of the Human IgG2 Subclass. *J. Biol. Chem.* **2008**, *23*, 16206–16215. [CrossRef] [PubMed]
14. Sattler, M.; Schmidt, P.; Schedletzky, O.; Glaser, S.J.; Sørensen, O.W.; Griesinger., C. A general enhancement scheme in heteronuclear multidimensional NMR employing pulsed field gradients. *J. Biomol. NMR* **1994**, *4*, 301–306.
15. Smallcombe, S.H.; Patt, S.L.; Keifer, P.A. WET solvent suppression and its applications to LC NMR and high-resolution NMR spectroscopy. *J. Magn. Reson. A.* **1995**, *117*, 295–303. [CrossRef]
16. Hwang, T.-L.; van Zijl, P.C.M.; Garwood, M. Asymmetric adiabatic pulses for NH selection. *J. Magn. Reson.* **1999**, *138*, 173–177. [CrossRef] [PubMed]

molecules

MDPI

Article

NMR Spectroscopy for Protein Higher Order Structure Similarity Assessment in Formulated Drug Products

Deyun Wang [1], You Zhuo [2], Mike Karfunkle [3], Sharadrao M. Patil [2], Cameron J. Smith [4], David A. Keire [5] and Kang Chen [2,*]

1 Northeast Medical Products Laboratory, Office of Regulatory Science, Office of Regulatory Affairs, U.S. Food and Drug Administration, Jamaica, NY 11433, USA; deyun.wang@fda.hhs.gov
2 Division of Complex Drug Analysis, Office of Testing and Research, Office of Pharmaceutical Quality, Center for Drug Evaluation and Research, U.S. Food and Drug Administration, Silver Spring, MD 20993, USA; you.zhuo@fda.hhs.gov (Y.Z.); sharadmpatil@gmail.com (S.M.P.)
3 Division of Pharmaceutical Analysis, Office of Testing and Research, Office of Pharmaceutical Quality, Center for Drug Evaluation and Research, U.S. Food and Drug Administration, St. Louis, MO 63110, USA; mike.karfunkle@fda.hhs.gov
4 Division of Liquid Based Products I, Office of Lifecycle Drug Products, Office of Pharmaceutical Quality, Center for Drug Evaluation and Research, U.S. Food and Drug Administration, Silver Spring, MD 20993, USA; cameron.smith@fda.hhs.gov
5 Office of Testing and Research, Office of Pharmaceutical Quality, Center for Drug Evaluation and Research, U.S. Food and Drug Administration, St. Louis, MO 63110, USA; david.keire@fda.hhs.gov
* Correspondence: kang.chen@fda.hhs.gov; Tel.: +1-240-402-5550

check for **updates**

Citation: Wang, D.; Zhuo, Y.; Karfunkle, M.; Patil, S.M.; Smith, C.J.; Keire, D.A.; Chen, K. NMR Spectroscopy for Protein Higher Order Structure Similarity Assessment in Formulated Drug Products. *Molecules* **2021**, *26*, 4251. https://doi.org/10.3390/molecules26144251

Academic Editor: Robert Brinson

Received: 21 May 2021
Accepted: 8 July 2021
Published: 13 July 2021

Publisher's Note: MDPI stays neutral with regard to jurisdictional claims in published maps and institutional affiliations.

Abstract: Peptide and protein drug molecules fold into higher order structures (HOS) in formulation and these folded structures are often critical for drug efficacy and safety. Generic or biosimilar drug products (DPs) need to show similar HOS to the reference product. The solution NMR spectroscopy is a non-invasive, chemically and structurally specific analytical method that is ideal for characterizing protein therapeutics in formulation. However, only limited NMR studies have been performed directly on marketed DPs and questions remain on how to quantitively define similarity. Here, NMR spectra were collected on marketed peptide and protein DPs, including calcitonin-salmon, liraglutide, teriparatide, exenatide, insulin glargine and rituximab. The 1D ^{1}H spectral pattern readily revealed protein HOS heterogeneity, exchange and oligomerization in the different formulations. Principal component analysis (PCA) applied to two rituximab DPs showed consistent results with the previously demonstrated similarity metrics of Mahalanobis distance (D_M) of 3.3. The 2D ^{1}H-^{13}C HSQC spectral comparison of insulin glargine DPs provided similarity metrics for chemical shift difference ($\Delta\delta$) and methyl peak profile, i.e., 4 ppb for ^{1}H, 15 ppb for ^{13}C and 98% peaks with equivalent peak height. Finally, 2D ^{1}H-^{15}N sofast HMQC was demonstrated as a sensitive method for comparison of small protein HOS. The application of NMR procedures and chemometric analysis on therapeutic proteins offer quantitative similarity assessments of DPs with practically achievable similarity metrics.

Keywords: similarity metrics; Mahalanobis distance; chemical shift difference; peak profile; relative peak height

1. Introduction

Complex generic and biosimilar drug products (DPs) are increasingly developed and comprehensive analysis of these DPs is the foundation for their regulatory approval [1–4]. The active pharmaceutical ingredient (API) or drug substance (DS) in protein DPs ranges in size from short peptides to large monoclonal antibodies (mAbs). The native folding of proteins, heterogeneity, dynamic exchange between conformations, oligomerization and aggregation profile in a formulation are collectively called the higher order structure (HOS) properties of protein therapeutics and are typically critical for efficacy and safety [5].

Protein HOS is stabilized by weak hydrogen bonding, electrostatic and hydrophobic forces, which are solvent dependent, and, consequently, formulation differences affect HOS [6–9]. In addition, proteins can be chemically modified either purposely, e.g., pegylation, or unintentionally, e.g., oxidation, which could introduce variability to protein HOS [10–12]. All these factors and the accompanying sensitivity to solution conditions necessitate characterizing the protein chemistry and HOS with minimal perturbation to the formulation by ideally using DPs [13]. In addition, the analytical means to assess protein HOS in a formulation are desired for generic and biosimilar drug developers that mostly only have access to the marketed originator DPs that are usually deemed as the reference DPs.

With the development of higher field strength magnets and cryogenic probes, modern high-resolution NMR spectroscopy is a non-invasive and sensitive method for protein molecular structure characterization [14–18]. However, several assumptions among stakeholders have limited the application of NMR on formulated DPs. The first is that strong excipient signals in a DP would interfere with weak DS signals such that NMR spectra would be dominated by the peaks of the excipients and would not be useful for protein HOS assessment. Indeed, NMR for protein HOS characterizations [19] has been applied on proteins extracted from DP [20–22], proprietary DS or non-marketed DP [23–25], which aimed to demonstrate the applicability of modern heteronuclear NMR to characterize proteins with ^{15}N and ^{13}C nuclei at natural abundances.

Second, the lack of acceptable metrics for similarity assessment means that most comparisons have been made at a visual level. The question of the level of similarity that is practically measurable remains to be answered quantitively. Previous attempts were made to collect NMR spectra on a DS enriched formulation of filgrastim [26] and DP formulations of insulin [27]. A combined chemical shift difference of 8 ppb or less was proposed as the threshold for experimental precision in 2D-NMR comparisons of biosimilars using data between the US and Indian marketed filgrastim DPs [26]. The principal component analysis (PCA) of the insulin DP NMR spectra revealed the practically achievable similarity threshold expressed in Mahalanobis distance (D_M) to be 3.3 or less [27]. These values were achieved when 600 MHz spectrometer with room temperature probe was used, therefore, the derived metrics were practical (in terms of the availability of instruments) and could be useful in establishing the acceptance criteria for a certain DP before and after a manufacturing change and for the comparison between a generic or biosimilar protein and the reference DP. However, their validity has not been further tested.

Third, the type of HOS properties reliably measured from DPs using modern NMR is not entirely clear. Herein, 1D ^1H NMR spectra were acquired on a range of marketed DPs with protein molecular weights ranging from 3 kDa to 145 kDa and with the protein concentration as low as 0.01 mM. The protein HOS properties of folding, intermediate exchange and oligomerization were all reflected in the NMR spectral patterns. Using rituximab DPs, the proposed D_M similarity metric was verified again. Using insulin glargine DPs, the methyl peak profile method showed that both chemical shift and relative peak height can be used to derive practically achievable similarity metrics. Finally, the sensitive ^1H-^{15}N sofast HMQC experiment was demonstrated to be a valuable NMR method to characterize the protein backbone HOS.

2. Results

The peptide and protein drug products (DPs) listed in Table 1 were sourced from the US market except Reditux®, which was sourced from India. All DPs are the reference drugs except Basaglar® and Reditux®, which are follow-on products to Lantus® and Rituxan®, respectively. All 1D ^1H, 2D ^1H-^{13}C and 2D ^1H-^{15}N NMR spectra were collected on formulated DPs with minimal dilution of adding 5% D_2O (v/v).

Table 1. Drug Products studied.

Drug Product	Drug Substance	Number of Amino Acids	M.W. (kDa)	Concentration (mM)	pH
Miacalcin®	Calcitonin-Salmon	32	3.43	0.0097 [1]	n/a
Saxenda®	Liraglutide	38	3.75	1.6	8.15
Forteo®	Teriparatide	34	4.12	0.061	4
Byetta®	Exenatide	39	4.19	0.060	4.5
HumulinR®	Insulin Human	51	5.81	0.60 [2]	n/a
Humalog®	Insulin Lispro	51	5.81	0.60 [2]	7.0–7.8
Lantus®Basaglar®	Insulin Glargine	53	6.06	0.60	4
Rituxan®Reditux®	Rituximab	1328	145	0.069	6.5

[1] Based on the equivalence between 1 mg and 6000 I.U. per USP NF.; [2] based on the equivalence between 0.0347 mg and 1 USP unit per USP NF and Eu. Pharm.; n/a: not available from the drug label.

2.1. 1D ^{1}H NMR Spectroscopy

2.1.1. Excipients

Excipients in protein formulations can function as preservatives (e.g., phenol and *m*-cresol), tonicity agents (e.g., mannitol), pH buffering agents (e.g., acetate) or protein stabilizers (e.g., polysorbate 80) [28]. The excipients are mostly small molecules at high concentration relative to the API. Due to the fast tumbling of excipients in liquid formulations, excipient peaks generally were sharper and more intense in the NMR spectra (Figure 1, left panels). Most excipient peaks were located in the high field region between 1 and 4 ppm, while preservatives with aromatic moieties had peaks between 6.5 and 7.5 ppm. The peaks were readily assignable with the help of chemical shift databases [29,30] or 2D ^{1}H-^{13}C spectra. The excipient polysorbate 80 (PS80) had a more complicated spectrum, with major peaks at 3.7, 2.3, 2.0, 1.6, 1.3 and 0.9 ppm [31]. Importantly, all excipient peaks should be excluded when protein HOS comparison is performed.

2.1.2. Process-Related Impurities

Small sharp peaks from process related impurities such as residual solvents and leachable were also identified and should be blinded out of HOS comparison as well [32,33]. For example, silicone oil used as a lubricant in DP containers could leach into the formulation and appear as a broad peak of polydimethylsiloxane (s) at 0.05 ppm, which can be further hydrolyzed to dimethylsilanediol and trimethylsilanol and appears as sharp peaks at 0.15 ppm (d) and 0.13 ppm (t), respectively (Figure 1A,C,D, right panels) [34]. Notably, the proton resonances of larger proteins could overlap with the spectral region around 0 ppm. Therefore, if NMR is used for the quantification of silicone oil components in protein formulations, T_2-filtered CPMG pulse train may be used to remove protein resonances [35,36].

2.1.3. Protein HOS

The protein DS may be formulated at concentration of about 1 mM or less (Table 1). The 1D ^{1}H NMR spectrum is the most sensitive NMR method to characterize protein HOS in DP formulations. The spectra need to be vertically enlarged by 2–4 orders of magnitude in order to visualize the lower intensity protein peaks (Figure 1, right panels). Among the tested DPs, the 3.43 kDa calcitonin-salmon is formulated at the lowest concentration of 9.7 µM. Calcitonin's sharp and dispersed amide peaks suggested that calcitonin-salmon adopts a folded monomeric HOS in formulation (Figure 1A). The 3.75 kDa liraglutide has a similar M.W. to calcitonin-salmon, however, broadened amide peaks were observed in liraglutide's spectra (Figure 1B), suggesting oligomerization of the protein in formulation.

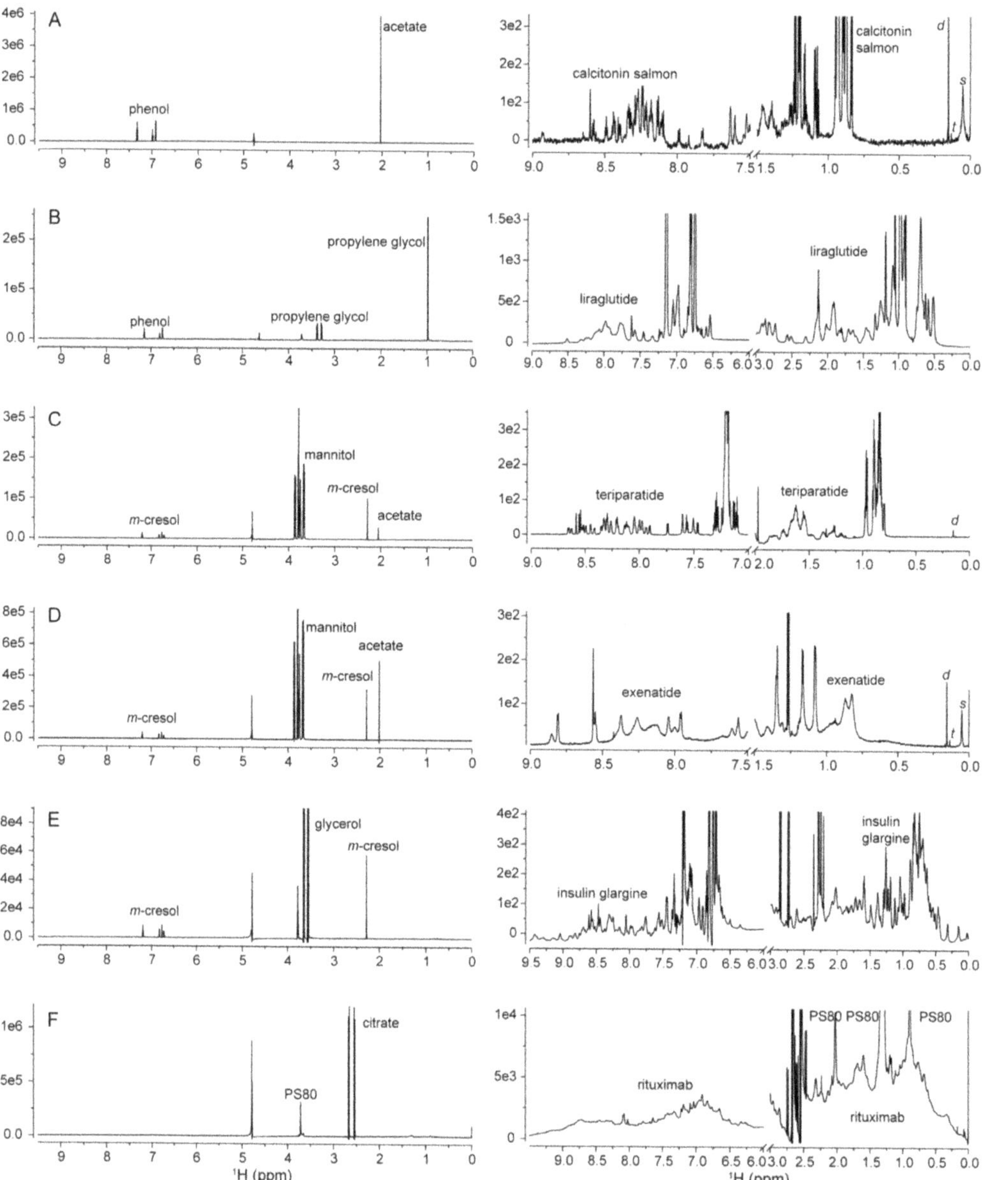

Figure 1. The 1D ^{1}H NMR spectra of protein drug products of Miacalcin® (**A**), Saxenda® (**B**), Forteo® (**C**), Byetta® (**D**), Lantus® (**E**) and Rituxan® (**F**) collected using an 850 MHz spectrometer. The spectra on the left are in full scale and those on the right are vertically enlarged and horizontally cut to display protein peaks. Signals from major excipients, drug substances and the leachable compounds dimethylsilanediol (d), trimethylsilanol (t) and polydimethylsiloxane (s) are annotated.

The 4.12 kDa teriparatide had sharp and dispersed amide peaks, suggesting a folded HOS in the formulation (Figure 1C). For the 4.19 kDa exenatide, much broadened peaks were observed while the detected number of peaks was much less (Figure 1D), suggesting the peptide was undergoing intermediate exchange broadening [37]. The observed exchange broadening is associated with exenatide in equilibrium between several HOS states and the exchange kinetics occur over a similar time scale of the chemical shift difference between different states, usually in the range of µs-ms exchange.

For the 6.06 kDa insulin glargine, the detected dispersed peaks suggest well folded HOS in the formulation at pH 4 (Figure 1E). Finally, the observed broadened peaks of the 145 kDa rituximab were due to its large M.W., but the dispersed amide peaks suggest the monoclonal antibody has a folded HOS (Figure 1F).

2.1.4. Similarity Metrics of D_M

Although the 1D ^{1}H spectra can be used to assess protein HOS qualitatively, a quantifiable similarity metric is of interest to demonstrate comparability after manufacturing changes or similarity between any two drug brands [38]. Previously, 1D ^{1}H spectra between the reference insulin and the follow-on insulin DPs were chemometrically compared using principal component analysis (PCA) and Mahalanobis distance (D_M) metrics, which were derived from PCA space [27]. The previous results on insulin DPs suggested a D_M value of 3.3 as the similarity threshold [27], where above 3.3 value there were clear differences in the HOS, while below that there were no discernable differences. Here, the rituximab DPs Rituxan® and Reditux® were compared using the same approach. Excipient free regions of the spectra including the down field amide spectrum are informative for rituximab HOS.

The 1D ^{1}H spectra of seven lots of Rituxan® and three lots of Reditux® (Table S1) were collected using both 850 MHz and 600 MHz spectrometers. Representative spectra from both brands were superimposed and visually appeared similar (Figure 2, left). Spectral regions belonging to excipients, residual solvents, water and blank were excluded before PCA. The PCA results showed that the first two principal components accounted for over 70% of the spectral intensity variations (Figure 2, right). The Mahalanobis distance (D_M) calculated from the first three principal components (Table S2) using Equations (1) and (2) were 1.95 and 3.15, when the 850 MHz data and the 600 MHz data were used, respectively. Both values were below the established similarity threshold D_M value of 3.3 [27], suggesting similar HOS between the two products. Ninety percent (90%) confidence interval ellipses were drawn for the Rituxan® DP lots in PC1/2 space (Figure 2, right). For the 850 MHz data, two lots of Reditux® fell outside the ellipse; for the 600 MHz data, one lot of Reditux® fell outside the ellipse. Therefore, the slight difference in field dependent D_M values is not necessarily correlated to the apparent differences shown in the ellipse circles. The results suggest any D_M values below the metric of 3.3 would indicate high similarity.

2.2. 2D ^{1}H-^{13}C Spectroscopy

2.2.1. Protein Specificity

As an alternative to 1D ^{1}H spectra, 2D ^{1}H-^{13}C heteronuclear single quantum coherence (HSQC) spectra are highly specific to protein sequence and HOS. The HSQC spectrum of the methyl chemical shift region, which was 12–27 ppm along the ^{13}C axis and 0–1.5 ppm along the ^{1}H axis, has higher sensitivity due to the 3 C-H correlations and fast internal rotational dynamics of the methyl group. In addition, major excipient or solvent peaks, if observed in a methyl HSQC spectrum, can be readily identified because of their strong intensity and unique ^{13}C chemical shifts, e.g., ethanol in insulin DPs of HumulinR®, Humalog® and Basaglar® (Figure 3). Therefore, methyl-HSQC spectra can be an ideal high-resolution spectrum for HOS assessment.

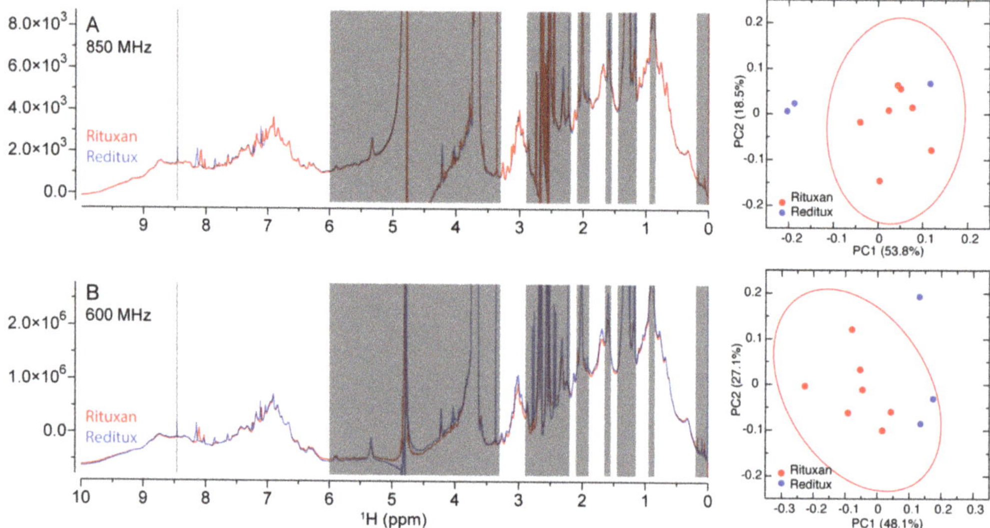

Figure 2. The superimposed 1D ^{1}H NMR spectra of representative rituximab drug products (DP) of Rituxan® and Reditux® collected using 850 MHz (**A**, left) and 600 MHz (**B**, left) spectrometers. The spectral regions in gray of 0–0.2 ppm, 0.85–0.95 ppm, 1.15–1.45 ppm, 1.55–1.65 ppm, 1.9–2.1 ppm, 2.2–2.9 ppm, 3.3–6 ppm and 8.45–8.47 ppm were excluded before principal component analysis (PCA). The resulting PCA scores for each DP lot from both 850 MHz data (**A**, right) and 600 MHz data (**B**, right) were plotted along the PC1 and PC2 axes. The 90% confidence ellipses are drawn for Rituxan® lots only (A/B, right).

The amino acids possessing one or two methyl groups are Alanine (Ala), Methionine (Met), Threonine (Thr), Valine (Val), Leucine (Leu) and Isoleucine (Ile). For insulin, methyl peaks of Ala, Thr, Val, Leu and Ile observed in the DP spectra of HumulinR® and Humalog® (Table 1) can be approximately assigned by referencing the literature assignment for insulin human [39] and insulin lispro [40] (Figure 3A). Due to formulation or buffer differences, the assignment can be confidently made for residues of Ile at positions A2 and A10, Ala at position B14 and Thr at positions A8, B27 and B30; ambiguities remain for the Val and Leu clusters in the ^{13}C chemical shift ranges of 21–22 ppm and 23–27 ppm, respectively (Figure 3A). Nevertheless, large chemical shift differences were observed between insulin human and insulin lispro, both of which have an identical M.W. of 5808 Da. The insulin lispro sequence differs from insulin human by two amino acid residues at positions B28–B29, Pro-Lys in insulin human and Lys-Pro in insulin lispro. Thus, the sequence difference introduced large changes in chemical shifts for almost every methyl peak (Figure 3A), which is consistent with a large change of HOS in the formulated API arising from only a two amino acids swap.

Shown in Figure 3B is the superimposed spectra between Lantus® and Basaglar®, both of which contain the same DS insulin glargine. The chemical shift assignment was not transferrable from other insulins because the spectra are significantly different (Figure 3A,B). The methyl peaks of insulin glargine were labeled with possible amino acid type and alphabetic letters. The total number of identified peaks ($s/n > 10$) was 48, which is more than the expected 28 methyl peaks calculated from the insulin glargine sequence. The increased peak number is attributed to some of the methyl groups adopting at least two slow exchange conformations in the formulation, e.g., Ala(B14) had two peaks of Ala-a and Ala-b at ^{13}C chemical shift of 19 ppm (Figure 3B). Overall, the methyl HSQC spectra between the two DPs are highly similar, suggesting that insulin glargine is folded in similar HOS for the two formulations.

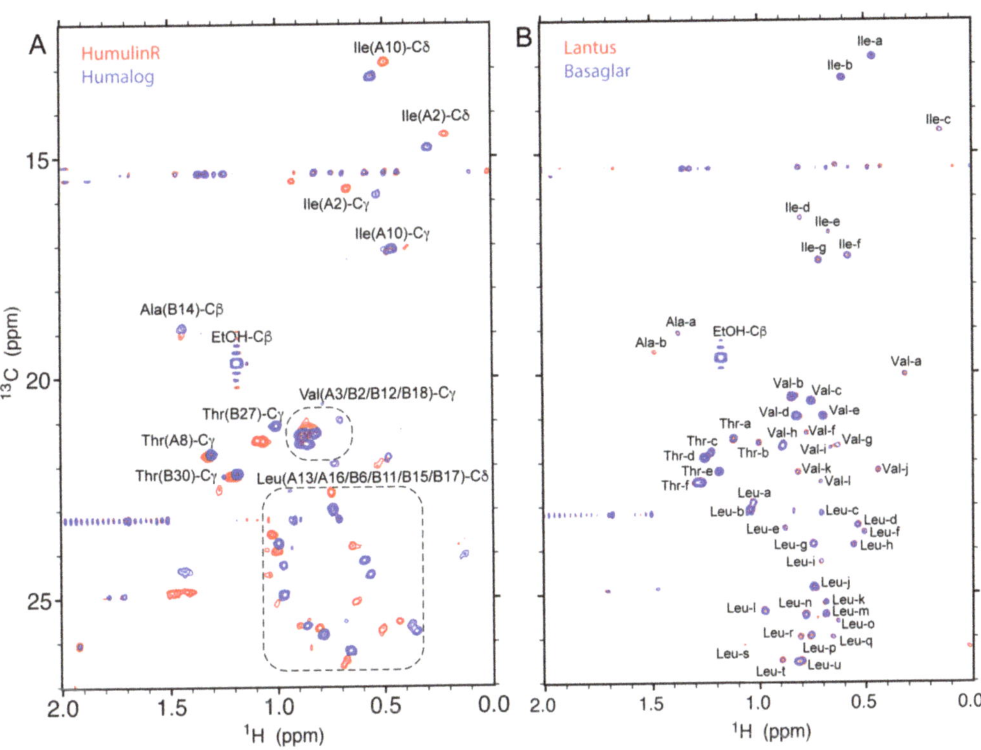

Figure 3. The superimposed 2D ^{1}H-^{13}C HSQC NMR spectra between insulin drug products of HumulinR® and Humalog® (**A**) and between Lantus® and Basaglar® (**B**) collected using a 600 MHz spectrometer. The plotting threshold of intensity was at a signal to noise ratio of 5 and 10 for (**A**) and (**B**), respectively. The methyl spectra of insulin human and insulin lispro were approximately assigned according to the published assignments; ambiguities were observed in the Valine and Leucine clusters shown in dashed boxes (**A**). The spectra of insulin glargine cannot be definitively assigned due to a large change in the observed chemical shifts and each peak was labeled for possible amino acid type and with an alphabetic letter (**B**).

2.2.2. Similarity Metrics of Δδ

Each peak in a 2D ^{1}H-^{13}C NMR spectrum has three dimensions, including peak intensity and ^{1}H and ^{13}C chemical shifts, all of which are sensitive to protein HOS. Previous spectral comparisons on insulin [38] and filgrastim [26] 2D spectra have applied PCA for similarity evaluation, which took into account all spectral variables from the three dimensions (two frequencies and intensity) for comparison. However, no similarity metrics were derived. The filgrastim ^{1}H-^{15}N spectral comparison established a combined chemical shift difference (CCSD) metric of 8 ppb [26]. The chemical shift comparison was repeated here for the 48 methyl peaks between Lantus® and Basaglar®. For each brand the inter-lot averaged chemical shift values were used as DP specific δ. The differences of chemical shift (Δδ) between the two DPs were plotted along both ^{1}H and ^{13}C axis (Figure 4A,B). The maximum ^{1}H Δδ was 3.4 ppb identified in the Leu-d peak. The maximum ^{13}C Δδ was −13 ppb identified in the Leu-j peak. When a 10% larger difference is permitted in the maximum Δδ, similarity metrics with rounded values of 4 and 15 ppb for the ^{1}H and ^{13}C chemical shifts, respectively, can be proposed. These metrics are on par with the previous CCSD metric of 8 ppb [26] or 4 ppb [41], which was a normalized Δδ value from both the ^{1}H and ^{15}N axes.

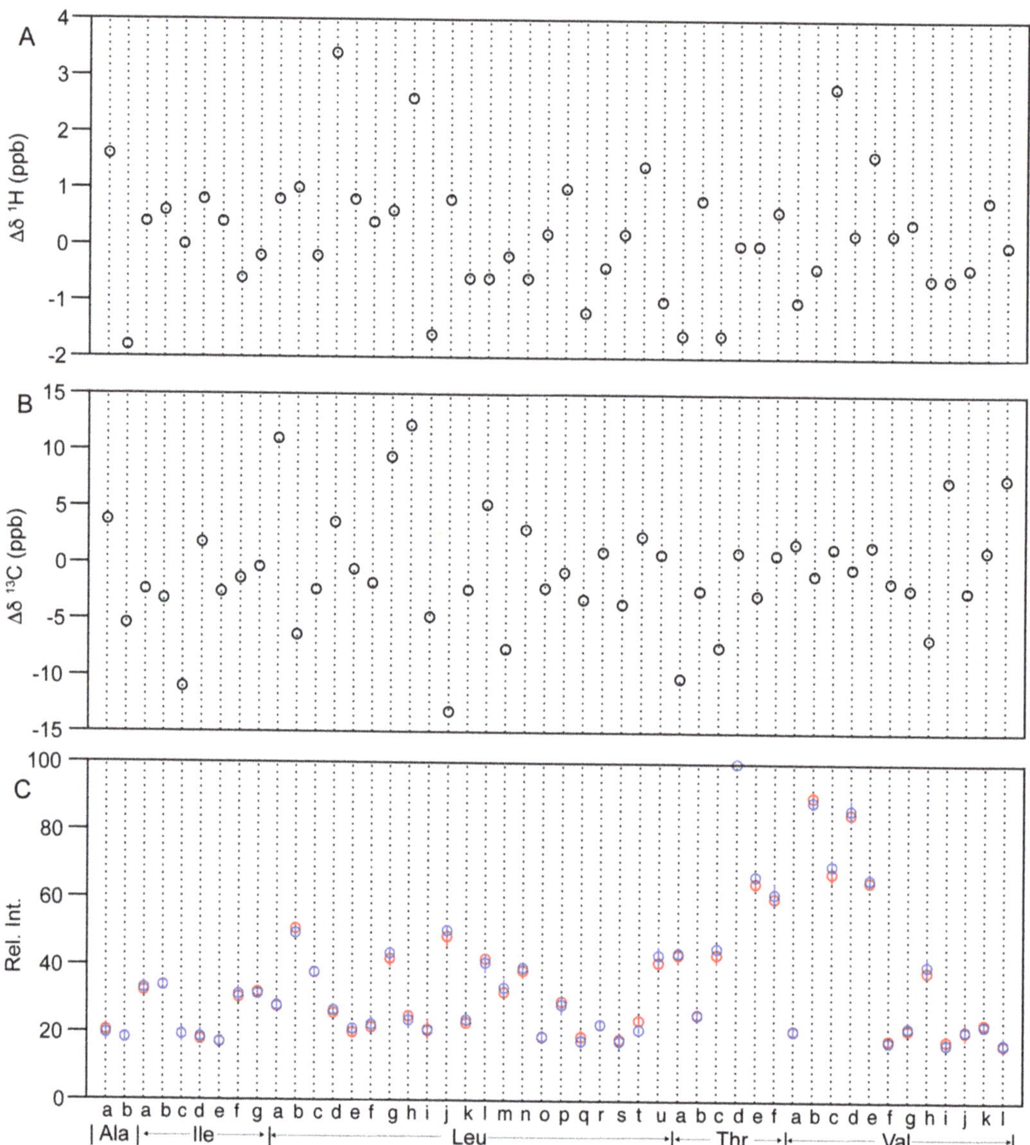

Figure 4. The chemical shift and relative peak height difference between insulin glargine drug products of Lantus® and Basaglar®. The ¹H (**A**) and ¹³C (**B**) chemical shift difference and the relative peak heights (**C**) were plotted along the labeled peaks of Figure 3B.

2.2.3. Methyl Peak Profile

The peak intensity was compared using peak heights. First, the absolute peak heights of the strongest peak, Thr-d, were tabulated for five lots of each brand and five technical repeats from one lot of Lantus® (Table S3). The calculated p value between the five technical repeats and the five lots of Lantus® was 0.35, demonstrating the technical issues related to the spectral differences were within the inter-lot DP differences. By contrast, the Thr-d peak height in Lantus® inter-lot spectra was on average 4% higher than the peak height of the Basaglar® inter-lot spectra. The 4% difference was significant with a p value of 0.0061 (Table S3), which is less than the threshold value of 0.05. The 4% difference may be related to differences in assay and response Q-factor of the NMR probe to different formulations, usually related to electric capacity or ionic strength [42]. As a result, the comparison using absolute peak height for all methyl peaks was deemed not suitable.

However, the relative peak heights, related to the dynamics and exchange kinetics of each methyl group should still be a valid choice for comparison purposes. Here, the ratios of each peak height to the Thr-d were calculated according to Equation (3). The mean and standard deviation from both Lantus® inter-lot spectra and Basaglar® inter-lot spectra were plotted in Figure 4C. The p values were calculated for all 48 peaks (Table 2) and 47 p values were higher than 0.05 except for the Leu-t peak with p value of 0.0055. Ultimately, 47 out of 48 peaks were equivalent in relative peak height between the two brands, demonstrating that the HOS distribution and exchange kinetics of the insulin glargine in the two DPs were similar. The work suggested the similarity metrics for peaks that show comparable relative peak height could be at least 98% (47/48).

2.3. 2D ^{1}H-^{15}N Spectroscopy

The 2D ^{1}H-^{15}N spectrum may be a more specific NMR experiment than ^{1}H-^{13}C HSQC to evaluate protein HOS because the amide ^{1}H and ^{15}N chemical shifts are exclusively sensitive to peptide backbone conformation. However, the ^{1}H-^{15}N HSQC experiment is at least one order of magnitude less sensitive than the methyl ^{1}H-^{13}C HSQC experiment. Thus, ^{15}N spectra via indirect detection in formulated DP samples can be challenging to collect when the DS concentration is less than 1 mM. The previously developed NMR pulse sequence of sofast-heteronuclear multi-quantum correlation (HMQC) has the advantage of shorter recycle delay without perturbing water resonances [43]. The sofast-HMQC experiment allows the 2D ^{1}H-^{15}N correlation spectrum to be collected within 24 h for DPs with protein concentrations as low as 0.06 mM. Representative spectra of Forteo®, Byetta® and Lantus® are shown in Figure 5. Amide peaks of the protein backbone and Asn/Gln side chains are observed in the ^{15}N chemical shift range of 108–129 ppm and the ^{1}H chemical shift range of 7.4–9.1 ppm. The number of detected peaks for teriparatide in Forteo® was 29 (Figure 5A), while a total of 44 peaks are expected. The 66% coverage suggests the teriparatide adopts a well-defined HOS in the formulation, which is consistent with the 1D ^{1}H spectral pattern (Figure 1C). By contrast, only six peaks were detected for exenatide in Byetta® (Figure 5B), whereas a total of 39 peaks are expected. The 15% coverage suggests the exenatide resonances are in intermediate exchange between different HOS forms, which is, again, consistent with the broadening in the 1D ^{1}H spectrum (Figure 1D). For insulin glargine in Lantus®, the 54 peaks detected account for 87% of the expected 62 peaks (Figure 5C). The results for insulin glargine suggest the existence of a single or fast averaged backbone HOS in the formulation.

Table 2. The relative peak height comparison of insulin glargine DPs.

Peak	Antus®					Basaglar®					*p* Value
	Lot 1	Lot 2	Lot 3	Lot 4	Lot 5	Lot 1	Lot 2	Lot 3	Lot 4	Lot 5	
Ala-a	17.5	20.7	19.8	20.8	23.4	20.2	17.0	22.7	20.8	17.6	0.61
Ala-b	18.0	19.0	17.4	18.1	19.5	16.3	18.8	17.3	18.7	20.7	0.96
Ile-a	30.2	34.7	32.7	30.6	32.8	34.5	33.2	29.6	35.5	32.4	0.54
Ile-b	35.3	34.5	33.6	32.5	33.3	34.9	34.3	32.2	34.1	33.3	0.92
Ile-c	20.1	21.9	19.2	17.8	17.4	20.9	21.1	15.5	18.3	21.0	0.96
Ile-d	20.0	16.8	17.8	16.9	18.0	19.8	20.8	17.2	15.2	20.2	0.57
Ile-e	16.9	15.9	16.4	17.3	19.2	19.3	15.7	16.5	16.7	17.9	0.93
Ile-f	29.8	32.3	27.4	30.4	31.9	30.6	31.4	35.0	28.7	30.7	0.52
Ile-g	32.0	31.4	30.6	30.9	34.6	31.0	32.6	31.8	31.0	30.4	0.54
Leu-a	27.6	26.4	26.7	27.2	30.7	29.1	27.9	25.4	26.6	30.8	0.84
Leu-b	52.8	52.0	48.8	49.4	50.7	50.9	49.3	50.3	46.5	50.3	0.28
Leu-c	36.9	37.6	36.3	39.3	38.4	39.9	35.9	38.0	38.3	37.2	0.88
Leu-d	25.7	25.9	26.7	24.8	25.8	26.6	25.3	27.0	27.0	27.2	0.11
Leu-e	22.0	17.7	18.8	20.9	21.1	23.4	20.9	22.3	19.4	20.0	0.35
Leu-f	22.5	18.8	21.6	21.2	24.0	23.6	25.7	22.1	18.5	22.3	0.58
Leu-g	41.1	41.1	39.6	41.7	46.2	44.7	43.6	43.5	40.8	45.7	0.26
Leu-h	23.3	26.6	23.6	25.6	25.8	25.9	24.6	21.2	21.6	25.7	0.35
Leu-i	20.1	22.2	16.9	21.1	24.8	21.3	21.0	21.1	20.2	19.9	0.83
Leu-j	46.4	50.1	44.5	49.9	52.0	51.8	50.8	48.3	49.6	51.1	0.28
Leu-k	21.7	25.4	21.7	23.1	23.0	26.2	22.1	23.7	24.9	22.6	0.38
Leu-l	42.4	41.8	41.6	41.3	42.3	43.9	41.4	37.5	38.0	42.9	0.43
Leu-m	31.2	31.0	32.7	31.2	33.9	34.2	34.6	29.1	32.9	36.0	0.34
Leu-n	36.2	38.2	37.2	38.4	42.3	40.8	37.6	37.3	40.5	40.5	0.52
Leu-o	18.1	17.6	18.8	21.4	18.5	18.8	17.3	20.2	21.6	17.7	0.85
Leu-p	30.6	30.6	29.4	28.0	28.1	31.7	28.7	24.7	26.4	30.8	0.57
Leu-q	19.1	21.8	19.7	16.1	18.3	18.6	19.9	15.1	16.7	18.7	0.38
Leu-r	24.2	22.7	23.0	21.9	21.5	22.6	24.7	23.5	20.5	22.2	0.96
Leu-s	15.3	19.1	18.4	17.3	21.4	15.7	17.9	15.6	18.9	21.0	0.75
Leu-t	23.0	25.3	23.7	25.3	22.4	21.7	21.5	21.6	19.6	21.3	0.0055
Leu-u	42.5	38.8	40.1	41.6	42.2	43.2	40.2	42.2	42.8	47.5	0.17
Thr-a	45.8	45.8	41.2	40.9	42.8	43.8	41.9	44.8	42.4	46.0	0.73
Thr-b	27.2	27.0	24.0	25.0	25.0	24.1	25.8	26.1	25.9	25.5	0.83
Thr-c	39.4	45.2	42.5	43.8	46.1	45.9	44.2	47.3	44.7	44.3	0.21
Thr-d	100	100	100	100	100	100	100	100	100	100	n/a
Thr-e	64.0	65.5	61.5	64.8	67.4	68.5	68.0	64.5	65.5	66.9	0.13
Thr-f	60.2	61.1	59.0	58.8	61.9	65.1	64.5	58.8	58.1	60.7	0.46
Val-a	22.5	21.6	22.1	19.1	20.0	22.3	21.7	18.7	20.6	21.3	0.85
Val-b	92.1	91.1	87.1	88.6	91.4	91.2	86.6	87.4	88.4	89.6	0.29
Val-c	67.2	65.7	64.6	68.9	71.8	73.0	70.7	67.1	69.2	69.6	0.2
Val-d	84.8	84.8	82.9	87.1	86.3	89.3	90.4	83.1	85.1	83.9	0.5
Val-e	64.2	65.2	63.6	67.1	65.6	68.1	67.2	62.5	66.4	65.3	0.52
Val-f	20.0	17.8	19.3	18.0	17.4	18.4	17.2	17.3	19.2	17.2	0.34
Val-g	20.9	20.2	21.8	21.3	23.5	20.9	21.6	21.1	22.1	25.1	0.52
Val-h	40.0	37.7	37.8	37.6	39.6	43.3	36.4	40.0	38.3	43.3	0.29
Val-i	15.9	20.7	17.5	17.3	20.0	18.1	19.3	16.4	15.4	17.4	0.42
Val-j	24.0	20.8	21.9	16.8	21.7	22.9	23.2	20.9	20.3	19.4	0.83
Val-k	24.2	21.1	25.7	22.8	23.0	21.5	21.9	24.0	24.3	22.0	0.55
Val-l	15.8	19.9	14.5	17.9	17.7	20.2	13.5	17.1	18.6	18.0	0.84

n/a: not available.

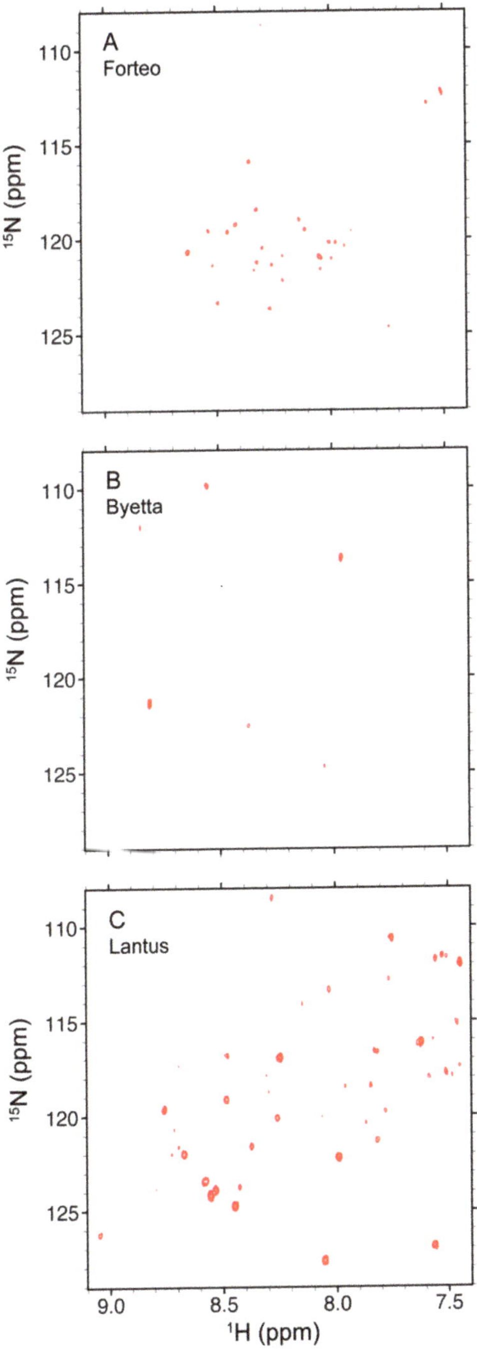

Figure 5. The 2D sofast ^{1}H-^{15}N HMQC NMR spectra of teriparatide in Forteo® (**A**), exenatide in Byetta® (**B**) and insulin glargine in Lantus® (**C**) collected using an 850 MHz spectrometer. The plotting threshold of intensity was at a signal to noise ratio of 5.

3. Materials and Methods

3.1. Drug Product NMR Samples

All the drug products (DP) listed in Table 1 were sourced from the US market except Reditux®, which was sourced from the India market. The DPs used for similarity metrics calculations were 7 lots of Rituxan®, 3 lots of Reditux®, 5 lots of Lantus® and 5 lots of Basaglar® (Table S1). NMR samples were prepared by directly mixing 0.5 mL of DP formulation with 0.03 mL of deuterium oxide, which contained 0.002% of trimethylsilylpropanoic acid (TMSP) or trimethylsilylpropanesulfonate sodium (DSS), then transferring to a 5 mm NMR precision tube (Wilmad-LabGlass).

3.2. NMR Spectrsocopy

All the NMR spectra were collected at experimental temperature of 25 °C. The NMR spectrometers were either a Bruker (Billerica, MA, USA) 850 MHz equipped with a cryogenic QCI probe or a Bruker 600 MHz equipped with a liquid nitrogen-cooled prodigy TCI probe.

3.2.1. 1D ^{1}H NMR Spectra Collection and Processing

The 1D ^{1}H NMR spectra shown in Figures 1 and 2A were collected using an 850 MHz spectrometer. The pulse program *p3919gp* was applied. The ^{1}H carrier was placed on the water resonance at 4.8 ppm. The spectral width was 14 ppm and a total of 23,808 complex points were collected. The acquisition time was 1 s and recycle delay was 2 s. The number of scans were 1024 for calcitonin-salmon and rituximab DPs, 256 for exenatide DP and 128 for liraglutide, teriparatide and insulin glargine DPs. Each free induction decay (FID) was apodized with a 90° shifted sine-square window function, scaled half for the first point, zero-order phase corrected and zero filled to a spectral size of 32k points before Fourier transform (FT). A baseline correction method of splines was applied for the calcitonin-salmon and teriparatide spectra and no correction was applied for the liraglutide, exenatide, insulin glargine and rituximab spectra. All the 1D NMR data were processed and analyzed using MestReNova 14.1 software (Mestrelab Research S.L.).

The 1D ^{1}H NMR spectra shown in Figure 2B were collected using a 600 MHz spectrometer. The pulse program of modified 1D NOESY *noe-p3919.kc* was applied [20]. The ^{1}H carrier was placed on the water resonance at 4.8 ppm. The spectral width was 13 ppm and a total of 16,384 complex points were collected. The acquisition time was 1 s and the recycle delay was 2 s. The NOE mixing time was 0.1 s. The number of scans was 1024. The NMR samples and data processing were identical to that used for the 850 MHz spectra.

3.2.2. 2D ^{1}H-^{13}C NMR Spectra Collection and Processing

The 2D ^{1}H-^{13}C HSQC spectra shown in Figure 3 were collected using a 600 MHz spectrometer. A modified sensitivity enhanced gradient HSQC pulse sequence *hsqcetgpsi2.kc* was applied [44]. The spectral width for the ^{1}H dimension was 11 ppm with the carrier frequency centered at 4.8 ppm. The spectral width for the ^{13}C dimension was 50 ppm with the carrier frequency centered at 23 ppm. The complex points of 1024 and 600 were acquired for the ^{1}H and ^{13}C dimensions, respectively. The resulting acquisition times for ^{1}H and ^{13}C spins were 78 and 40 ms, respectively. The ^{13}C decoupling sequence was GARP with a radio frequency field strength of 1.9 kHz. The coupling constant $^1J_{HC}$ was set to 155 Hz as a compromise between efficient INEPT transfer and T_2 signal loss. The recycle delay was 2 s. The number of scans was 16 and the total experimental time was 6 h.

The data processing was performed using NMRPipe [45]. The apodization function of cosine was applied to both dimensions of ^{1}H and ^{13}C. The first point was scaled with a factor of 0.5 before zero-order phase correction. Zero filling of 2048 × 1024 real data points was applied to the ^{1}H and ^{13}C dimensions. The baseline corrections on frequency domains were carried out with a polynomial function under auto mode. The chemical shift reference followed the established procedure [46]. HSQC peaks with s/n higher than 10 were picked and peak heights were recorded using Sparky (Sparky 3, UCSF).

3.2.3. 2D ^{1}H-^{15}N NMR Spectra Collection and Processing

The sofast 2D ^{1}H-^{15}N HMQC spectra shown in Figure 5 were collected using an 850 MHz spectrometer. The Bruker pulse sequence of *sfhmqcf3gpph* was applied. The spectral width for the ^{1}H dimension was 14 ppm with the carrier frequency centered at 4.8 ppm. The spectral width for the ^{15}N dimension was 35 ppm with the carrier frequency centered at 117 ppm. The complex points of 1784 and 200 were acquired for the ^{1}H and ^{15}N dimensions, respectively. The resulting acquisition times for ^{1}H and ^{15}N spins were 75 and 33 ms, respectively. The ^{15}N decoupling sequence was GARP with a radio frequency field strength of 1.1 kHz. The coupling constant $^1J_{HN}$ was set to 100 Hz as a compromise between efficient INEPT transfer and T_2 signal loss. The recycle delay was 0.1 s. The number of scans was 2000 and the total experimental time was 23 h. The ^{1}H-^{15}N spectra were processed in a manner similar to the ^{1}H-^{13}C spectra, except for the zero filling of 4096 × 1024 real data points applied to the ^{1}H and ^{15}N dimensions.

3.3. Calculation of Similarity Metrics

The similarity metrics were calculated using the above processed 1D ^{1}H spectra of Rituxan® and Reditux® and 2D ^{1}H-^{13}C spectra of Lantus® and Basaglar®.

3.3.1. Mahalanobis Distance (D_M) between 1D Spectra

The 1D ^{1}H NMR spectra of Rituxan® and Reditux® were used to calculate Mahalanobis distance [47]. The procedure was described previously [27]. Briefly, principal component analysis (PCA) was performed on the spectra of 7 lots of Rituxan® and 3 lots of Reditux®. The spectral regions corresponding to peaks of excipient and solvent were excluded, including regions of 0–0.2 ppm, 0.85–0.95 ppm, 1.15–1.45 ppm, 1.55–1.65 ppm, 1.9–2.1 ppm, 2.2–2.9 ppm, 3.3–6 ppm and 8.45–8.47. The rest of the spectra were binned at 0.01 ppm resolution, resulting in a total of 370 bins with summed spectral intensities within each bin. The summed intensities were subject to integrity checks, sum normalization and Pareto scaling before PCA using MestReNova 14.1 (Mestrelab Research S.L.).

The Mahalanobis distances (D_M) between the two rituximab DPs were calculated using PC1-3 scores. PCA scores from all the lots of each brand were tabulated as sample matrices of $A_{m \times p}$ and $B_{n \times p}$ for Rituxan® and Reditux®, respectively, with m or n representing the number of lots and p representing the number of principal components used toward D_M calculation. In the present study, m was 7 for Rituxan®, n was 3 for Reditux® and p was 3. The mean vector $\overline{A}_{1 \times p}$ and covariance matrix $S_{A,p \times p}$ was calculated using Rituxan® sample matrix $A_{m \times p}$. In parallel, the mean vector $\overline{B}_{1 \times p}$ and covariance matrix $S_{B,p \times p}$ was calculated using Reditux® sample matrix $B_{n \times p}$. The covariance matrices of the two were averaged per Equation (1) before calculating D_M using Equation (2). The calculations were performed using MATLAB 9.0 (The MathWorks Inc.) and the code can be found in the Supplementary Materials.

$$S = (mS_A + nS_B)/(m+n) \tag{1}$$

$$D_M = \sqrt{(\overline{A} - \overline{B})S^{-1}(\overline{A} - \overline{B})\prime} \tag{2}$$

3.3.2. Chemical Shift Difference ($\Delta\delta$) between 2D Spectra

The 2D ^{1}H-^{13}C NMR spectra of Lantus® and Basaglar® were used to calculate the chemical shift difference ($\Delta\delta$). A total of 48 methyl peaks were identified with a signal to noise (s/n) ratio over 10. The peaks were approximately assigned to amino acid residue types of Ala, Thr, Ile, Leu and Val. Within each amino acid residue type, the peaks were labeled with alphabetic letters. The ^{1}H and ^{13}C chemical shift of each brand were averaged from the spectra of the 5 lots, representing the mean chemical shift of the peak in each brand. Chemical shift difference ($\Delta\delta$) was the difference between the mean values of Basaglar® and Lantus®.

3.3.3. Methyl Peak Profile between 2D Spectra

The peak heights of the 48 identified methyl peaks in 2D ^{1}H-^{13}C spectra of Lantus$^®$ and Basaglar$^®$ were recorded as I_x for peak x. The relative peak intensity of the peak x ($Rel.Int._x$) was calculated per Equation (3), where I_{Thr-d} is the peak height of the peak Thr-d, which is the peak with the highest intensity. The mean and standard deviation of $Rel.Int._x$ were calculated from the spectra of 5 different lots within each brand (Table 2). The p value was calculated for each peak x using t-test function of two-sample assuming unequal variances in Excel (ver. 16.46). The significant threshold of 0.05 in p value was used to determine the equivalence of relative peak heights. The equation for relative peak intensity is described as follows.

$$Rel.Int._x = 100 \times I_x / I_{Thr-d} \tag{3}$$

4. Discussion and Conclusions

4.1. HOS Inferrred from 1D and 2D Spectra

In this work, standard NMR experiments using 1D ^{1}H, 2D ^{1}H-^{13}C HSQC and 2D ^{1}H-^{15}N sofast HMQC pulse sequences were performed on formulated protein DPs. The NMR peak patterns from both 1D and 2D spectra are qualitatively informative for protein HOS properties, providing insight into the oligomerization of liraglutide, the HOS exchange of exenatide and the well folded HOS of calcitonin-salmon, teriparatide, insulin glargine and rituximab. In general, the 1D ^{1}H NMR experiment provides information on the HOS profile and whether a protein is folded in formulation. Information on more specific HOS variation can be obtained from heteronuclear 2D spectra. Each 2D spectrum was sensitive to different aspect of HOS. For example, in the insulin glargine spectra, while the methyl ^{1}H-^{13}C spectrum showed the sidechains adopting two slowly exchanging conformers, the ^{1}H-^{15}N spectrum was more consistent with a single well-folded backbone conformer. The two observations were not necessarily inconsistent with each other, rather, they illustrate the complex nature of protein HOS in the formulation and the atomic level probes used by the different NMR experiments.

4.2. HOS Similarity Metrics Calculated from 1D and 2D Spectra

What is different from the pioneering work on demonstrating heteronuclear 2D NMR at protein natural abundance [41,48,49] is that the current study uses NMR on formulated DPs and also includes deriving practically achievable similarity metrics. Earlier work demonstrated the practically achievable Mahalanobis distance (D_M) value of 3.3 based on the PCA of 1D ^{1}H spectra collected on the marketed insulin reference product and follow-on products [27]. Here, we obtained the D_M values of 1.95 and 3.15 using PCA and 1D ^{1}H spectra of rituximab DPs marketed in the US and India, suggesting that a D_M metrics value of less than 3.3 could be a general acceptance criterion.

While PCA can be conveniently performed on 1D spectra and has been demonstrated on 2D spectra [26,38,50], PCA is challenging to implement for 2D spectra because of the technical complications in binning the 2D spectra and avoiding non-DS peaks at the same time. An alternative method is to focus on the DS peak profile. The normalized distance comparison approach was proposed to compare 2D spectra along the axes of chemical shifts and peak intensity; however, no acceptance criteria were ever proposed [51]. Here the previous chemical shift comparison method [26] was verified using 2D ^{1}H-^{13}C spectra collected on insulin glargine DPs and the chemical shift different metrics ($\Delta\delta$) of 4 ppb for ^{1}H and 15 ppb for ^{13}C were derived. Furthermore, the peak profile method [44] was adopted to compare the relative peak heights between two insulin glargine brands, where p values were derived from t-test. In these insulin spectra, 98% of the methyl cross peaks had equivalent relative peak heights between the two brands. These 2D spectral similarity metrics could be equivalent to the D_M value of 1.6 obtained by using 1D spectra [27]. The methyl peak profile results represent another practically achievable similarity metrics for 2D spectral comparison.

In summary, the NMR data collected in the current study provided examples of simple experiments and analyses on formulated protein DP and demonstrated practical measurements to assess equivalence of HOS between different DPs. The metrics proposed were validated using marketed similar DPs that were manufactured differently and are proposed as a benchmark to determine the degree of similarity for protein HOS in formulated DPs.

Supplementary Materials: The following are available online. Tables S1–S3 and Matlab code for D_M calculation.

Author Contributions: Conceptualization, D.A.K. and K.C.; methodology, D.W., D.A.K. and K.C.; validation, D.W. and Y.Z.; investigation, D.W., Y.Z., M.K., S.M.P. and K.C.; writing—original draft preparation, D.W., C.J.S., D.A.K. and K.C.; writing—review and editing, D.W., M.K., S.M.P., C.J.S., D.A.K. and K.C. All authors have read and agreed to the published version of the manuscript.

Funding: This research received no external funding.

Institutional Review Board Statement: Not applicable.

Informed Consent Statement: Not applicable.

Data Availability Statement: The data presented in this study are available in supplementary material.

Acknowledgments: This project was supported, in part, by an appointment (Y.Z. and S.M.P.) to the Research Participation Program at the CDER administered by the Oak Ridge Institute for Science and Education (ORISE) through an interagency agreement between the U.S. Department of Energy and the U.S. FDA.

Conflicts of Interest: The authors declare no conflict of interest.

Disclaimer: This article reflects the views of the author and should not be construed to represent U.S. FDA's views or policies.

Sample Availability: Samples of the compounds are not available from the authors.

References

1. Zhang, L.; Lionberger, R.A. Generics 2030: Where Are We Heading in 2030 for Generic Drug Science, Research, and Regulation? *Clin. Pharmacol. Ther.* **2020**, *107*, 1293–1295. [CrossRef] [PubMed]
2. Christl, L.A.; Woodcock, J.; Kozlowski, S. Biosimilars: The US Regulatory Framework. *Annu. Rev. Med.* **2017**, *68*, 243–254. [CrossRef] [PubMed]
3. Fisher, A.C.; Lee, S.L.; Harris, D.P.; Buhse, L.; Kozlowski, S.; Yu, L.; Kopcha, M.; Woodcock, J. Advancing pharmaceutical quality: An overview of science and research in the US FDA's Office of Pharmaceutical Quality. *Int. J. Pharm.* **2016**, *515*, 390–402. [CrossRef] [PubMed]
4. Kozlowski, S.; Woodcock, J.; Midthun, K.; Behrman Sherman, R. Developing the Nation's Biosimilars Program. *N. Engl. J. Med.* **2011**, *365*, 385–388. [CrossRef] [PubMed]
5. Moussa, E.M.; Panchal, J.P.; Moorthy, B.S.; Blum, J.S.; Joubert, M.K.; Narhi, L.O.; Topp, E.M. Immunogenicity of Therapeutic Protein Aggregates. *J. Pharm. Sci.* **2016**, *105*, 417–430. [CrossRef] [PubMed]
6. Gruia, F.; Du, J.; Santacroce, P.V.; Remmele, R.L.; Bee, J.S. Technical Decision Making with Higher Order Structure Data: Impact of a Formulation Change on the Higher Order Structure and Stability of a mAb. *J. Pharm. Sci.* **2015**, *104*, 1539–1542. [CrossRef] [PubMed]
7. Aubin, Y.; Hodgson, D.J.; Thach, W.B.; Gingras, G.; Sauvé, S. Monitoring Effects of Excipients, Formulation Parameters and Mutations on the High Order Structure of Filgrastim by NMR. *Pharm. Res.* **2015**, *32*, 3365–3375. [CrossRef] [PubMed]
8. Bramham, J.E.; Podmore, A.; Davies, S.A.; Golovanov, A.P. Comprehensive Assessment of Protein and Excipient Stability in Biopharmaceutical Formulations Using 1H NMR Spectroscopy. *Acs Pharmacol. Transl. Sci.* **2020**, *4*, 288–295.
9. Panjwani, N.; Hodgson, D.J.; Sauvé, S.; Aubin, Y. Assessment of the Effects of pH, Formulation and Deformulation on the Conformation of Interferon Alpha-2 by NMR. *J. Pharm. Sci.* **2010**, *99*, 3334–3342. [CrossRef]
10. Shah, D.D.; Singh, S.M.; Mallela, K.M.G. Effect of Chemical Oxidation on the Higher Order Structure, Stability, Aggregation, and Biological Function of Interferon Alpha-2a: Role of Local Structural Changes Detected by 2D NMR. *Pharm. Res.* **2018**, *35*, 17. [CrossRef]
11. Hodgson, D.J.; Aubin, Y. Assessment of the structure of pegylated-recombinant protein therapeutics by the NMR fingerprint assay. *J. Pharm. Biomed. Anal.* **2017**, *138*, 351–356. [CrossRef]
12. Casagrande, F.; Dégardin, K.; Ross, A. Protein NMR of biologicals: Analytical support for development and marketed products. *J. Biomol. NMR* **2020**, *74*, 657–671. [CrossRef]

13. Weiss, W.F.; Gabrielson, J.P.; Al-Azzam, W.; Chen, G.; Davis, D.L.; Das, T.K.; Hayes, D.B.; Houde, D.; Singh, S.K. Technical Decision Making with Higher Order Structure Data: Perspectives on Higher Order Structure Characterization from the Biopharmaceutical Industry. *J. Pharm. Sci.* **2016**, *105*, 3465–3470. [CrossRef]

14. Keire, D.A. Analytical Tools for Physicochemical Characterization and Fingerprinting. In *Science and Regulations of Naturally Derived Complex Drugs*; Sasisekharan, R., Lee, S.L., Rosenberg, A., Walker, L.A., Eds.; Springer International Publishing Ag: Cham, Switzerland, 2019; Volume 32, pp. 91–113.

15. Guerrini, M.; Rudd, T.R.; Yates, E.A. NMR in the Characterization of Complex Mixture Drugs. In *Science and Regulations of Naturally Derived Complex Drugs*; Sasisekharan, R., Lee, S.L., Rosenberg, A., Walker, L.A., Eds.; Springer International Publishing Ag: Cham, Switzerland, 2019; Volume 32, pp. 115–137.

16. Amezcua, C.A.; Szabo, C.M. Assessment of Higher Order Structure Comparability in Therapeutic Proteins Using Nuclear Magnetic Resonance Spectroscopy. *J. Pharm. Sci.* **2013**, *102*, 1724–1733. [CrossRef]

17. Arbogast, L.W.; Brinson, R.G.; Formolo, T.; Hoopes, J.T.; Marino, J.P. 2D 1HN, 15N Correlated NMR Methods at Natural Abundance for Obtaining Structural Maps and Statistical Comparability of Monoclonal Antibodies. *Pharm. Res.* **2016**, *33*, 462–475. [CrossRef]

18. Arbogast, L.W.; Brinson, R.G.; Marino, J.P. Mapping Monoclonal Antibody Structure by 2D 13C NMR at Natural Abundance. *Anal. Chem.* **2015**, *87*, 3556–3561. [CrossRef]

19. Kiss, R.; Fizil, Á.; Szántay, C. What NMR can do in the biopharmaceutical industry. *J. Pharm. Biomed. Anal.* **2018**, *147*, 367–377. [CrossRef]

20. Chen, K.; Long, D.S.; Lute, S.C.; Levy, M.J.; Brorson, K.A.; Keire, D.A. Simple NMR methods for evaluating higher order structures of monoclonal antibody therapeutics with quinary structure. *J. Pharm. Biomed. Anal.* **2016**, *128*, 398–407. [CrossRef]

21. Hodgson, D.J.; Ghasriani, H.; Aubin, Y. Assessment of the higher order structure of Humira®, Remicade®, Avastin®, Rituxan®, Herceptin®, and Enbrel® by 2D-NMR fingerprinting. *J. Pharm. Biomed. Anal.* **2019**, *163*, 144–152. [CrossRef]

22. Japelj, B.; Ilc, G.; Marušič, J.; Senčar, J.; Kuzman, D.; Plavec, J. Biosimilar structural comparability assessment by NMR: From small proteins to monoclonal antibodies. *Sci. Rep.* **2016**, *6*, 32201. [CrossRef]

23. Elliott, K.W.; Ghasriani, H.; Wikström, M.; Giddens, J.P.; Aubin, Y.; Delaglio, F.; Marino, J.P.; Arbogast, L.W. Comparative Analysis of One-Dimensional Protein Fingerprint by Line Shape Enhancement and Two-Dimensional 1H,13C Methyl NMR Methods for Characterization of the Higher Order Structure of IgG1 Monoclonal Antibodies. *Anal. Chem.* **2020**, *92*, 6366–6373. [CrossRef] [PubMed]

24. Franks, J.; Glushka, J.N.; Jones, M.T.; Live, D.H.; Zou, Q.; Prestegard, J.H. Spin Diffusion Editing for Structural Fingerprints of Therapeutic Antibodies. *Anal. Chem.* **2016**, *88*, 1320–1327. [CrossRef] [PubMed]

25. Poppe, L.; Jordan, J.B.; Lawson, K.; Jerums, M.; Apostol, I.; Schnier, P.D. Profiling Formulated Monoclonal Antibodies by 1H NMR Spectroscopy. *Anal. Chem.* **2013**, *85*, 9623–9629. [CrossRef] [PubMed]

26. Ghasriani, H.; Hodgson, D.J.; Brinson, R.G.; McEwen, I.; Buhse, L.F.; Kozlowski, S.; Marino, J.P.; Aubin, Y.; Keire, D.A. Precision and robustness of 2D-NMR for structure assessment of filgrastim biosimilars. *Nat. Biotechnol.* **2016**, *34*, 139. [CrossRef]

27. Wang, D.; Park, J.; Patil, S.M.; Smith, C.J.; Leazer, J.L., Jr.; Keire, D.A.; Chen, K. An NMR-Based Similarity Metric for Higher Order Structure Quality Assessment Among, U.S. Marketed Insulin Therapeutics. *J. Pharm. Sci.* **2020**, *109*, 1519–1528. [CrossRef]

28. Ionova, Y.; Wilson, L. Biologic excipients: Importance of clinical awareness of inactive ingredients. *PLoS ONE* **2020**, *15*, e0235076. [CrossRef]

29. Wishart, D.S.; Knox, C.; Guo, A.C.; Eisner, R.; Young, N.; Gautam, B.; Hau, D.D.; Psychogios, N.; Dong, E.; Bouatra, S.; et al. HMDB: A knowledgebase for the human metabolome. *Nucleic Acids Res.* **2009**, *37*, 603–610. [CrossRef]

30. Ulrich, E.L.; Akutsu, H.; Doreleijers, J.F.; Harano, Y.; Ioannidis, Y.E.; Lin, J.; Livny, M.; Mading, S.; Maziuk, D.; Miller, Z.; et al. BioMagResBank. *Nucleic Acids Res.* **2008**, *36*, 402–408. [CrossRef]

31. Patil, S.M.; Li, V.; Peng, J.; Kozak, D.; Xu, J.; Cai, B.; Keire, D.A.; Chen, K. A Simple and Noninvasive DOSY NMR Method for Droplet Size Measurement of Intact Oil-In-Water Emulsion Drug Products. *J. Pharm. Sci.* **2019**, *108*, 815–820. [CrossRef]

32. Wang, W.; Ignatius, A.A.; Thakkar, S.V. Impact of Residual Impurities and Contaminants on Protein Stability. *J. Pharm. Sci.* **2014**, *103*, 1315–1330. [CrossRef]

33. Gottlieb, H.E.; Kotlyar, V.; Nudelman, A. NMR Chemical Shifts of Common Laboratory Solvents as Trace Impurities. *J. Org. Chem.* **1997**, *62*, 7512–7515. [CrossRef]

34. Malmstrøm, J. Quantification of Silicone Oil and Its Degradation Products in Aqueous Pharmaceutical Formulations by 1H-NMR Spectroscopy. *J. Pharm. Sci.* **2019**, *108*, 1512–1520. [CrossRef]

35. Suh, M.S.; Patil, S.M.; Kozak, D.; Pang, E.; Choi, S.; Jiang, X.; Rodriguez, J.D.; Keire, D.A.; Chen, K. An NMR Protocol for In Vitro Paclitaxel Release from an Albumin-Bound Nanoparticle Formulation. *AAPS Pharmscitech* **2020**, *21*, 136. [CrossRef]

36. Skidmore, K.; Hewitt, D.; Kao, Y.-H. Quantitation and characterization of process impurities and extractables in protein-containing solutions using proton NMR as a general tool. *Biotechnol. Prog.* **2012**, *28*, 1526–1533. [CrossRef]

37. Palmer, A.G.; Kroenke, C.D.; Patrick Loria, J. Nuclear Magnetic Resonance Methods for Quantifying Microsecond-to-Millisecond Motions in Biological Macromolecules. In *Methods in Enzymology*; James, T.L., Dötsch, V., Schmitz, U., Eds.; Academic Press: Cambridge, MA, USA, 2001; Volume 339, pp. 204–238.

38. Chen, K.; Park, J.; Li, F.; Patil, S.M.; Keire, D.A. Chemometric Methods to Quantify 1D and 2D NMR Spectral Differences among Similar Protein Therapeutics. *AAPS Pharmscitech* **2018**, *19*, 1011–1019. [CrossRef]

39. Chang, X.; Jørgensen, A.M.M.; Bardrum, P.; Led, J.J. Solution Structures of the R6 Human Insulin Hexamer. *Biochemistry* **1997**, *36*, 9409–9422. [CrossRef]
40. Quinternet, M.; Starck, J.P.; Delsuc, M.A.; Kieffer, B. Heteronuclear NMR provides an accurate assessment of therapeutic insulin's quality. *J. Pharm. Biomed. Anal.* **2013**, *78*, 252–254. [CrossRef]
41. Brinson, R.G.; Marino, J.P.; Delaglio, F.; Arbogast, L.W.; Evans, R.M.; Kearsley, A.; Gingras, G.; Ghasriani, H.; Aubin, Y.; Pierens, G.K.; et al. Enabling adoption of 2D-NMR for the higher order structure assessment of monoclonal antibody therapeutics. *mAbs* **2019**, *11*, 94–105. [CrossRef]
42. Nishizaki, Y.; Lankin, D.C.; Chen, S.-N.; Pauli, G.F. Accurate and Precise External Calibration Enhances the Versatility of Quantitative NMR (qNMR). *Anal. Chem.* **2021**, *93*, 2733–2741. [CrossRef]
43. Schanda, P.; Brutscher, B. Very Fast Two-Dimensional NMR Spectroscopy for Real-Time Investigation of Dynamic Events in Proteins on the Time Scale of Seconds. *J. Am. Chem. Soc.* **2005**, *127*, 8014–8015. [CrossRef]
44. Zhuo, Y.; Keire, D.A.; Chen, K. Minor N-Glycan Mapping of Monoclonal Antibody Therapeutics Using Middle-Down NMR Spectroscopy. *Mol. Pharm.* **2021**, *18*, 441–450. [CrossRef] [PubMed]
45. Delaglio, F.; Grzesiek, S.; Vuister, G.W.; Zhu, G.; Pfeifer, J.; Bax, A. Nmrpipe-a Multidimensional Spectral Processing System Based on Unix Pipes. *J. Biomol. NMR* **1995**, *6*, 277–293. [CrossRef] [PubMed]
46. Wishart, D.S.; Bigam, C.G.; Yao, J.; Abildgaard, F.; Dyson, H.J.; Oldfield, E.; Markley, J.L.; Sykes, B.D. 1H, 13C and 15N chemical shift referencing in biomolecular NMR. *J. Biomol. NMR* **1995**, *6*, 135–140. [CrossRef] [PubMed]
47. Brereton, R.G. The Mahalanobis distance and its relationship to principal component scores. *J. Chemom.* **2015**, *29*, 143–145. [CrossRef]
48. Aubin, Y.; Gingras, G.; Sauvé, S. Assessment of the Three-Dimensional Structure of Recombinant Protein Therapeutics by NMR Fingerprinting: Demonstration on Recombinant Human Granulocyte Macrophage-Colony Stimulation Factor. *Anal. Chem.* **2008**, *80*, 2623–2627. [CrossRef]
49. Chen, K.; Freedberg, D.I.; Keire, D.A. NMR profiling of biomolecules at natural abundance using 2D 1H–15N and 1H–13C multiplicity-separated (MS) HSQC spectra. *J. Magn. Reson.* **2015**, *251*, 65–70. [CrossRef]
50. Arbogast, L.W.; Delaglio, F.; Schiel, J.E.; Marino, J.P. Multivariate Analysis of Two-Dimensional 1H, 13C Methyl NMR Spectra of Monoclonal Antibody Therapeutics To Facilitate Assessment of Higher Order Structure. *Anal. Chem.* **2017**, *89*, 11839–11845. [CrossRef]
51. Haxholm, G.W.; Petersen, B.O.; Malmstrøm, J. Higher-Order Structure Characterization of Pharmaceutical Proteins by 2D Nuclear Magnetic Resonance Methyl Fingerprinting. *J. Pharm. Sci.* **2019**, *108*, 3029–3035. [CrossRef]

MDPI
St. Alban-Anlage 66
4052 Basel
Switzerland
Tel. +41 61 683 77 34
Fax +41 61 302 89 18
www.mdpi.com

Molecules Editorial Office
E-mail: molecules@mdpi.com
www.mdpi.com/journal/molecules